THINKING OFF YOUR FEET

THINKING OFF YOUR FEET

HOW EMPIRICAL PSYCHOLOGY VINDICATES ARMCHAIR PHILOSOPHY

MICHAEL STREVENS

THE BELKNAP PRESS OF HARVARD UNIVERSITY PRESS

Cambridge, Massachusetts
London, England
2019

Copyright © 2019 by the President and Fellows of Harvard College
All rights reserved
Printed in the United States of America

First printing

Library of Congress Cataloging-in-Publication Data
Names: Strevens, Michael, author.
Title: Thinking off your feet : how empirical psychology vindicates armchair
 philosophy / Michael Strevens.
Description: Cambridge, Massachusetts : The Belknap Press of Harvard
 University Press, 2019. | Includes bibliographical references and index.
Identifiers: LCCN 2018014208 | ISBN 9780674986527 (alk. paper)
Subjects: LCSH: Analysis (Philosophy) | Empiricism. | Induction (Logic)
Classification: LCC B808.5 .S765 2018 | DDC 146/.4—dc23 LC record
available at https://lccn.loc.gov/2018014208

It is like what we imagine knowledge to be:
dark, salt, clear, moving, utterly free

—Elizabeth Bishop, "At the Fishhouses"

Contents

	List of Figures	*ix*
1	Philosophical Knowledge	1
2	Classical and Modern Conceptual Analysis	23
3	Other Forms of Conceptual Analysis	49
4	The Psychology of Philosophy	63
5	Natural Kind Concepts	75
6	Conceptual Inductivism	97
7	Inductivism versus Conceptual Analysis	117
8	Inductive Analysis	133
9	Reference	149
10	The Travails of Analysis	185
11	Against Essential Natures	211
12	Substance: Basic Natural Kinds	227
13	Substance: Philosophical Categories	251
14	Learning without the Senses	271
15	The Life and Death of Secondary Categories	295
	References	*327*
	Acknowledgments	*337*
	Index	*339*

List of Figures

Figure		page
5.1	Transformed raccoon	80
5.2	Essentialist theory of swans	84
5.3	Pre-chemistry essentialist theory of water	94
5.4	Chemically informed essentialist theory of water	95
6.1	Causal minimalist theories of water and swanhood	98
8.1	Conceptual analysis versus inductive analysis	137
12.1	Causal minimalist theory of swanhood: Two-step version	245
13.1	Basic natural kind concept topology generalized	253
13.2	Theories of justification	268
15.1	Theory of swans with single primary explainer	319

THINKING OFF YOUR FEET

CHAPTER 1

Philosophical Knowledge

1.1 Philosophical Knowledge

Philosophers, it is said, are people who sit in armchairs and gain substantive knowledge about such things as causality, moral responsibility, and the nature of objects simply by thinking. What these philosophers learn I call *philosophical knowledge*. Philosophical knowledge is distinguished, then, by three properties: a certain subject matter, which I will not characterize further, but which distinguishes philosophical knowledge from mathematical and (perhaps) phenomenological knowledge; a certain method, that of armchair reflection, about which I will have much more to say; and the property of being substantial rather than trivial or empty. I'll go no further than that in defining philosophical knowledge and allied notions: as Ingvar Kamprad has taught us, when assembling a complex structure, it is better not to tighten things to an exacting degree until all the parts are more or less in place.

Is there such a thing as philosophical knowledge? Perhaps not. Perhaps the claims that fill the philosophy journals and the great books fall far short of the standards for knowledge—perhaps they are mere speculation. If so, philosophers produce no knowledge whatsoever, and a fortiori, they produce no philosophical knowledge.

Then again, perhaps philosophers produce knowledge, but no philosophical knowledge, because the knowledge they do create lacks one of the three properties enumerated above.

Perhaps it is not obtained through reflection. Of course, the vast majority of philosophers' beliefs do come through reflection; perhaps, however, it is only the nonreflectively attained remainder—such as metaphysical conclusions delivered by scientific inquiry—that constitute knowledge.

Or perhaps philosophers' knowledge is not philosophical knowledge because it does not, despite appearances, concern philosophical subject matter. What looks to be knowledge about causality, for example, might merely be knowledge about normal human (or Western, or twenty-first century) patterns of causal thought—it might be psychological, rather than philosophical, knowledge.

Finally, perhaps what philosophers have succeeded in coming to know is not substantial. Perhaps, for example, what is learned by armchair reflection about causality is a set of necessary truths about causality, but necessary truths that are trivial consequences of the semantics of causal talk. They are not merely truths about semantics, but truths about causality itself—yet they lack philosophical substance, perhaps because they are not mind-independent, or because they do not genuinely constrain the way that the universe might be configured (Unger 2014).

These various worries are sufficiently disturbing that it would be better to start, not by considering ways in which we might barely fall short of reaching philosophical enlightenment, but by asking how anything like philosophical knowledge is possible at all. Thanks to what process could the mind imaginably arrive at some substantive knowledge of the nature of things without leaving the living room and emerging, bleary-eyed and blinking, into the real world?

This book aims to show how philosophical knowledge may be produced, when conditions are favorable, by one particular armchair method: philosophical analysis.

1.2 Philosophical Analysis

What makes something a chair? That it is designed for sitting, perhaps. But that is not enough: a couch is not a chair. That it is designed for one person to sit in, then. Still not enough: a stool is not a chair. Add to the list of

1.2. PHILOSOPHICAL ANALYSIS

requirements, in that case, that chairs should offer support for the back. You are not there yet: a car seat is not a chair. Append independent mobility to the list?

And so on. Any philosopher will recognize this process: a hypothesis as to the nature of some sort of thing is proposed, counterexamples are exhibited, the hypothesis revised to accommodate the counterexamples, again and again, for as many cycles as the journal page count, the needs of the profession, or the dialogue form can sustain. Some call this form of investigation the *method of cases*; I will call it (focusing on its particular subject matter) *philosophical analysis*.

Philosophical analysis is structurally similar to the process of hypothesis formation and testing in empirical inquiry—indeed, in science—but what plays the refuting or supporting role in analysis is delivered not by observation but by reflection, or more specifically, by judgments as to whether or not various specimens are instances of the thing to be analyzed, such as the judgment that recliners are chairs but couches, stools, and car seats are not. (What about the philosopher-king's throne?) For expository simplicity, let me assume that the thing to be analyzed is always a category rather than a property or a relation—for example, the category of morally good acts rather than the property of moral goodness—and call the judgments (or "intuitions") about category membership that furnish the evidence in philosophical analysis the *case judgments*. (Some intuitions are not in any significant sense case judgments: the intuition that time flows or the intuition that I am conscious or that my body is extended in space, for example. These I put aside; they play no role in this book, nor in philosophical analysis as it is usually understood, though they might surely play another kind of role in grounding philosophical knowledge.)

In the course of philosophical analysis as it is found in the wild, more is normally brought to bear on the investigation than the case judgments in isolation. A proposed analysis might be rejected because it is "implausible" or "baroque," because it fits badly with the best available analyses of related categories, or because it violates some global constraint such as physicalism. I will discuss the role of these considerations soon enough, but I begin by examining the role of case judgments in isolation—not because analysis based purely on case judgments predominates in philosophical thinking, but because it is a microcosm within which the problem of philosophical knowledge can be posed in a restricted and therefore simple, but nevertheless compelling, form.[1]

What the case judgments bear upon—the hypotheses—are also rather different in philosophical analysis than in scientific inquiry. The goal of philosophical analysis of any sort is to learn as much as possible about the fact or facts in virtue of which a specimen falls into the philosophical category in question—the facts in virtue of which something counts as knowledge, or as art, or as a causal connection. Let me call this fact or facts the category's *essential nature*. Philosophical analysis aims, then, to discover essential natures.

Such talk of natures is, I stipulate, quite compatible with nominalism or "conceptualism" about the category in question. I do not assume that essential natures are observer-independent or that they have a metaphysical status above and beyond our using them to organize the world in certain ways for our own convenience. For a category to have an essential nature, then, it need not correspond to a universal, or be in some respect "natural," or slice up nature's pliant carcass into anything that would be recognized by a competent butcher. What is required is simply that there is something in virtue of which the members of the category belong to the category. To describe that something—that essential nature—is what the philosophical analyst tries to do.

1.3 The Miracle of Armchair Knowledge

Philosophical analysis, unless our profession is deluding itself, can fruitfully be conducted in an armchair. Nothing goes in, but something new—philosophical knowledge—comes out. How can this be?

The miracle is accomplished, as I have said, in two steps. In the first, case judgments are made about particular specimens. Is this knowledge? Is that causation? In the second, a kind of hypothesis testing takes place, in which the facts delivered by the case judgments serve as evidence for or against various theses about a category's essential nature.

---

[1] On the strategy of focusing on this very simple version of analysis, compare Goldman and Pust (1998, 179).

1.3. THE MIRACLE OF ARMCHAIR KNOWLEDGE

That the second step should be feasible in the armchair is not so astonishing: even empirical scientists may bring their hard-won evidence to bear on a theory while equipped with nothing more than pencil, paper, and comfortable seating.

It is the first step that looks unbelievable in both senses of the word: while the scientist expends sometimes vast amounts of sweat and treasure to interrogate nature and elicit its most suggestive secrets—spending billions on particle accelerators and orbiting telescopes, or years administering surveys or wading through tropical swamps—the armchair philosopher simply lounges in the eponymous item of upholstery asking themselves straightforward questions as to what counts as what. The examination of philosophical analysis therefore ought to begin with an examination of the case judgments: their epistemic status, their scope, and their ultimate source.

In this section I lay out the properties of case judgments that need to be explained; in the next I will consider some historically important strategies for explaining them—or explaining them away.

First, epistemic status. If we are to acquire philosophical knowledge through analysis, case judgments ought to be for the most part reliable, warranted, and (we might hope) transparently so.

It seems that we get even more than we ask for. Case judgments are not only all of those things; they are, at least in some important instances, manifestly certain. When I decide that a stool is not a chair, I feel that my judgment is the last possible word on the matter. I simply could not be mistaken (assuming that I am thinking clearly) in ruling out couches, car seats, and so on as specimens of chairhood.

The same quality infuses certain philosophically more significant judgments. Sylvie throws a stool at the bookstore window; it breaks. Had the stool failed to shatter the window, Bruno would have broken it with a chair. Thus, even if Sylvie had not tossed her stool, the window would have broken. Yet the fact that the breaking of the window does not counterfactually depend on Sylvie's throw does not in the least disincline us to count her vandalistic act as the cause of the window's breaking. Nothing I could learn about the physics of causality would persuade me—so I feel—to abandon the judgment that the presence of a backup is irrelevant to an event's causal status.

I am equally convinced of my infallibility in judging a classic Gettier case (Gettier 1963). With her own eyes Sylvie sees Bruno filch *Twilight of the Idols* from the bookstore. She consequently entertains a justified belief that Bruno has in his possession Nietzsche's book. In fact, Bruno suffered an *untermenschlich* pang of guilt and replaced the book when Sylvie wasn't looking. Sylvie's belief is nevertheless true: there is a copy of the book in a box that Bruno is keeping for a friend, but which he has never opened. Sylvie does not know that Bruno possesses the book, I judge, in spite of her having a justified true belief to that effect. What could I possibly learn that would lead me to set aside this verdict? Not a thing. I call this phenomenon of felt certainty about judgments about cases—at least, about large numbers of cases that are interesting or central—*case certainty*.

I have described this certainty in two not-quite-equivalent ways. It might be that, on the one hand, in each of these cases I am sure that I have enough information to make a definitive judgment. Or it might be, on the other hand, that I am sure that no further information could undermine my judgment.

The second, weaker kind of certainty arises in the following way. I have information that disposes me to make a certain judgment. I may not at that stage feel certain about the judgment; I am merely quite confident. However, on further reflection I see that there is nothing more to know; I have all the information that is in principle relevant to deciding the case at hand (except of course the "answer" itself). My confidence, then, though not maximal, is unshakable.

In the latter case, "certainty" might be slightly too strong a word for the assurance I feel. For that matter, the word might seem too strong even in the former case for someone who has run philosophy's historical gauntlet from Pyrrho to Descartes to Hume to Nietzsche.[2] Regard *case certainty* as a term of art, then, that does not imply literal certainty, but something in the neighborhood.

---

[2] My case certainty about Gettier judgments, for example, does not rule out skeptical scenarios such as the possibility that I am delusional, subject to thought control, or sufficiently detached from the world that my thoughts fail to refer. Even at its strongest, it is certainty only relative to the usual suppositions under which thinking proceeds.

1.3. THE MIRACLE OF ARMCHAIR KNOWLEDGE

It is, of course, possible to think that this certainty, or high confidence, is horribly misplaced. To encourage such suspicions is the strategy of certain experimental philosophers whose doubts will be considered in due course; for the present, I assume that case certainty is to be taken at face value and is therefore to be explained rather than debunked.

Along with case certainty comes a sense that case judgments articulate necessary truths, or at least truths with wide modal scope of some sort. The impression of necessity arises, it seems, from the robustness of the judgments: I am prepared to judge that the justified true belief in a Gettier scenario falls short of qualifying as genuine knowledge in almost any context. Provided that justified true belief can exist at all, it can be Gettierized, and Gettierized, it is not knowledge.

Necessity may be too strong a notion to capture the scope of these truths. Williamson (2007) argues that the judgments are valid in the closest possible worlds in which the presuppositions and stipulations of a thought-experimental scenario hold, but not necessarily in every such possible world. Malmgren (2011) disagrees, arguing for necessity. I will not take a stand: this is one of many issues raised by analysis that, in the interest of brevity but at the expense of completeness, this book will overlook.

Yet another feature often attributed to the case judgments is their constituting a priori knowledge. The attribution may stem from the same source as the attribution of necessity—the judgments' independence from substantive suppositions—or perhaps from the impression of certainty or simply the nonempirical state of mind of the working armchair philosopher.

Throughout this book, I put aside questions about the modal force and the aprioricity of case judgments.[3] My questions will rather be: What reasons do we have to think that our case judgments (or at least some significant portion of them) are reliable? And why do we seem so sure?

These problems are difficult enough in themselves, but they become even more baffling when juxtaposed with the demand that philosophical knowledge be substantive. Let me explain.

A loose distinction may be made between concepts that are relatively open and those that are relatively closed. A *closed concept* has its boundaries determined in a way that is fixed more or less independently of the way

---

[3] The sole exception is the discussion of Chalmers's argument for aprioricity in Section 7.3.

the world is. An *open concept* has its boundaries determined in a way that depends to some extent on substantive matters of fact. It is, then, "open" in the way a mind may be open, taking into account context before fixing its reference.[4]

Rather than a definition, I offer a metaphor and an example. A closed concept may be compared to a stamp or die. It imposes its pattern on the world no matter what. An open concept is more like molten wax: it flows into the world and assumes a shape determined largely by what it finds there. We know the shape that will be punched out by the stamp before we bring it down on the material; to learn the shape that will be assumed by the wax, by contrast, we must know something of the worldly mold into which it is poured.

The concept "rare yellow ductile malleable metal" is relatively closed; whether something falls under the concept is decided by the possession of properties that are represented explicitly in the concept itself. The concept "metal with the same molecular structure as these ingots here" (gesturing at a local gold hoard) is relatively open; its instances are decided by a property that is only indirectly represented by the concept, and so that is in some sense left up to the world to fix. In both cases the concept's extension is (more or less) determinate, but in one case the rule that determines the extension depends to a far greater degree on the world than in the other.

Knowledge of the essential nature of the category picked out by the latter concept—which includes knowledge of the relevant molecular structure itself—seems to be substantive in a way that knowledge of the essential nature of the former category—a conjunction of observable properties—is not. And this fact is surely closely connected to the one's being relatively open and the other's being relatively closed.

Now, the problem. Case certainty concerning membership of a category suggests that the corresponding concept is closed: if it is possible to make a decisive judgment, in fullest confidence, as to whether a broad range of specimens belong to the category, independently of any substantive knowledge about the world, then the determiners of category membership must be fixed independently of substantive facts about the world. They must operate like the stamp that punches the same shape out of any material. But

---

[4] Compare Bealer's (1996, 23) distinction between semantically stable and unstable terms.

1.3. THE MIRACLE OF ARMCHAIR KNOWLEDGE

then it follows that knowledge of the category's essential nature is most likely not substantive. To constitute an object of substantive knowledge, an essential nature must be like the wax that assumes the shape of its surroundings, that comes to reflect some aspect of its domain. It might seem, then, that whatever knowledge of essential natures can be gained through armchair analysis is not substantive knowledge.

I intend this argument to be suggestive rather than demonstrative. (Indeed, I will do my best to find a way around it.) It has two weak points. The first is that I have given only a loose characterization of the open/closed distinction. But that is less important, I think, than the second, which is that I have not put my finger on the reason why knowledge of closed essential natures is typically not substantive.[5] Is it merely because closed essential natures are, because already incorporated into the concept, in some sense already known? That would be a relatively mild objection. Or is it because truths about closed essential natures are "conceptual," and so exist only in virtue of our concepts and not in virtue of the world? A third possibility is that truths about closed essential natures, even if they are not conceptual, are unlikely to reflect the fault lines of reality—"nature's joints"—since they articulate boundaries imposed on the world independent of (because closed to facts about) the objectively important boundaries.

As I develop explanations of case certainty based on various theories of concepts in the following chapters, I will be able to give these matters further consideration, extracting tailor-made arguments from what is, as presented here, more a sense of foreboding, a presentiment of insubstantiality, a nagging suspicion that, as Reichenbach (1951, 304) declared, "certainty is inseparable from nothingness."

Here, then, is a to-do list for any methodologist eager to account for the epistemology of philosophical analysis. (Alternatively, it may be regarded as a list of features that must be debunked or explained away by underminers of philosophical analysis.)

First, demonstrate or at least argue for the reliability of a large class of case judgments—large enough for analysts to determine the essential natures of a suitably wide range of philosophical categories. The reliability of

---

[5] By a "closed essential nature," I mean the essential nature of a category picked out by a closed concept, a formulation too tedious to spell out in the main text.

the case judgments is of course in itself much to be desired, but the demonstration also serves the purpose of providing warrant for our reliance on the judgments. (The question whether all analysts apprehend this warrant, or whether it is revealed only to a privileged meta-minded few, will be posed in the next chapter.) There may be other ways, besides an argument for reliability, to secure warrant for our dependence on case judgments, but reliability is the epistemic gold standard: anything else would give us cold comfort.

Second on the list is to explain case certainty: the very high confidence, and sense of immunity from defeat by further information, that we have for some of the philosophically most significant case judgments.

Third, the ability of case judgments to provide substantive knowledge ought to be defended. Especially welcome would be an explanation how substantiality can coexist with certainty—an explanation, in other words, why certainty about case judgments does not imply the insubstantiality of the conclusions about essential natures drawn from those judgments.

These three items are the principal challenges, I think, to a defense of analysis's power to produce philosophical knowledge, but several other desiderata ought to be kept in mind. For one, analysis should turn out to be, at least in many cases, genuinely an armchair process. It certainly looks the ticket. Then, the knowledge it produces ought to be of philosophical matters. Again, to all appearances, philosophical analysis meets this requirement. Finally, the inferences from case judgments to essential natures must be sound, and thus capable of licensing the conclusions that they appear to underwrite.

1.4 Empirically Informed Analysis

I have been writing so far as if all philosophical analysis is conducted in an empirical void. That is in one sense true by fiat: what happens in the armchair happens, by the very nature of the metaphor, in complete isolation from sensory input. But there is no philosophical prohibition on taking empirical knowledge into the armchair, that is, on conducting case-based philosophy in the light of previously acquired empirical knowledge.

Indeed, the satisfactory resolution of many philosophical questions appears to require or at least to thrive on empirical input:

1.4. EMPIRICALLY INFORMED ANALYSIS

Do genotypes "program" or "code for" phenotypes?
How is causality implemented in our universe?
To what extent do mental disorders diminish moral responsibility?
Is it permissible to switch off life support for a person in a persistent vegetative state?
What forms of government maximize human freedom?

I want to allow that inquiries into matters like these produce philosophical knowledge. Let me explain what I have in mind by examining an example that I hope to build upon prodigiously in the course of this book: the question "What is water?"—that is, the question of the essential nature of water.

To answer that question correctly requires empirical input: the chemists must get to work. But observation alone cannot resolve the issue; laboratory labor must be complemented by some concerted thinking that in both its method and its complications rather resembles philosophical analysis. Any talented chemist can experimentally determine that water is largely composed of H_2O, but from then on, the going gets rather more conceptual. Clearly a liquid need not be pure H_2O to qualify as a sample of water: seawater does not merely contain water, but is a kind of water itself. Is there a general cutoff point, a percentage of H_2O above which everything is water and below which nothing is? No; some impurities are more easily able to undermine waterhood than others: coffee, which is not a kind of water, contains more H_2O than seawater.

The inquiry into waterhood continues in Chapters 10 and 11; for now, I put it on hold. What I want you to see is that the "thinking phase" of the inquiry is much like philosophical analysis: it proceeds in the armchair, plainly using the method of cases to investigate the metaphysics of water. I have characterized philosophical knowledge as in some strict sense produced purely by armchair cogitation. In what sense, then, does empirically informed analysis, such as the inquiry into the nature of water, produce philosophical knowledge?

Think of what's learned in the armchair phase of inquiry as a conditional proposition relating a way the world could be to a fact about essential natures. In the case of water, the conditional says that if the world's chemistry works in such and such a way—as it happens, the actual way—then the essential nature of water is such and such.[6] That is artificial and abstract, of course—the typical analyst of water likely does not think of themself, in the course of their investigation, as learning such a conditional—but it draws the boundaries in the right place for the purpose of understanding philosophical methodology. In the case of water, then: in the empirical phase of inquiry, you learn the chemical facts, and in the armchair phase of inquiry, you learn what, given those chemical facts, the essential nature of water would be. Putting the two sets of knowledge together—the empirical and the philosophical knowledge—you learn the nature of water. What this picture makes clear is that even empirically informed philosophical analysis involves the acquisition of philosophical knowledge.

Sometimes it is only a very short step from the empirical knowledge that the philosopher takes with them into the armchair to their final conclusion. The philosophical knowledge acquired in the armchair in such cases is minimal or even trivial. (This might be true, for example, when a philosopher uses their knowledge of physics to answer the question "What is the nature of pressure in a gas?") Even when armchair investigation turns on a great deal of empirical knowledge, however, the philosophical knowledge involved may also be considerable. To decide under what circumstances mental illness lessens moral responsibility presumably requires much psychological research and also much armchair examination of the nature of responsibility.

That said, many of the paradigms of armchair philosophical investigation, such as the analysis of knowledge, rely on little empirical input. I don't presume that the analysis of knowledge is entirely free of empirical foundations. It may well depend, for example, on observationally acquired knowledge of belief/desire psychology. But by comparison with the analysis of water or diminished responsibility, the empirical input seems preliminary, a matter of stage setting, rather than a major part of the production itself.

To sum up, much philosophical analysis is empirically informed—perhaps all of it. We can nevertheless distinguish a part of what emerges

---

[6] The epistemic role I attribute to these world-to-essential-nature conditionals is inspired by David Chalmers's work on philosophical analysis (Chalmers 2012). Chalmers's own use of the conditionals—rather different from mine—is explored in Section 7.3.

from such investigations as conclusions reached by armchair thinking alone. These armchair products take the form of world-to-essential-nature conditionals, though I will for the most part avoid plaguing you with formulations in these terms. My aim in this book is to explain the origins and epistemic standing of any armchair thinking based on the method of cases, whether its conclusions are conditional or—as in the case of philosophical investigations that proceed quite free of empirical supposition—absolute. In so doing, I will take myself to have vindicated all philosophical analysis, to whatever degree it is empirically informed.

1.5 How Is Philosophical Knowledge Possible?

There are two great stories about philosophical knowledge. Here I tell them in loose and summary form; as such, they constitute grand strategies for accomplishing, or in some cases declining to accomplish, the items on the metaphilosopher's agenda.

According to an old story, philosophical reasoning is like any other kind of inquiry. There is a matter of fact, at first beyond our grasp, as to the nature of the philosophical categories. We begin with some clues. Thinking hard about these clues, we learn more and more about the categories—about the nature of knowledge and the good; about truth and the beautiful; about the causal, the nomological, the modal. Such thinking can go wrong, not only if we reason badly, but if the clues—the beliefs that comprise the starting point of our thinking—are mistaken. There is, however, some reason to think that the clues are correct.

On the nature of this reason, tellings of the old story diverge. Most august is the Platonic or Cartesian solution, on which the mind is incarnated with certain truths about the external world already in place. For Plato, these truths are acquired during a prenatal sojourn in the world of Forms (or something equally epistemically effective for which the *Phaedo*'s story stands); for Descartes they are implanted by a benevolent creator. In a more recent variation of Descartes's solution the well-meaning demiurge is Darwinian: natural selection has ensured that the human mind is adequately supplied with essential know-how and, more importantly, know-that.

These are nativist versions of the old story. To many moderns, the nativist explanation of the source of philosophical knowledge is laughably—or sadly—optimistic: there are no forms, no gods, and no reason to expect epistemic quarter from natural selection.

A non-nativist rendition of the old story might posit a faculty of philosophical intuition that is able to grasp directly, as though by perception, metaphysical or moral truths. Or it might opt for absolute idealism, on which knowledge of the external world is just a kind of self-knowledge, and so is in principle available by reflection. But I will give such ideas no further attention in this book, taking the nativist strain as the canonical telling of the old story.

According to a newer story, philosophical inquiry is quite different from ordinary inquiry into the nature of things, because the philosophical categories are human constructions, projections of the mind's own twisted associations onto the empirical plane. Philosophers' task, then, is to shine a light not out into the world but into the mind to illuminate the principles of construction, the rules by which the mind organizes its perception of or hypotheses about the world. Their method is consequently singular, peculiar, inward-looking, omphaloskeptic: transcendental philosophy or conceptual analysis. But executed correctly it cannot fail, because the object of inquiry has nowhere to hide. The new story is the Kantian story, the logical empiricist story, the story as told by the executors of the Canberra Plan.[7]

Whether recovery of the principles of construction supplies knowledge of the external world, let alone substantial knowledge, is difficult to say. The empiricist tendency is to deny that learning organizing principles confers knowledge of something outside the head; the constructivist inclines the other way. The empiricist is also happy to deny the substantiality of truths inferred from organizing principles; the constructivist is more reluctant, though they may eventually concede. In what follows, however, I will be happy to call all versions of the new story "constructivist" in a broad sense.

There are more stories than the old and the new. A contemporary contender holds that philosophical inquiry is continuous with scientific inquiry: the substantive knowledge that allows us to solve philosophical problems gets into the mind in the same way as all other substantive knowledge,

---

[7] The lead Canberra planner is perhaps Jackson (1998).

namely, through the senses (Quine 1960; related views may be found in Papineau [2009] and Paul [2012]). This might be called the scientific solution to the problem of philosophical knowledge—although perhaps, since it denies the existence of a special method of armchair inquiry, it is a dissolution.[8]

And then there is the skeptical story. The new story flirts with the idea that philosophical knowledge is not substantive: the scientific story with the idea that it is not accomplished solely in the armchair. The skeptical story is marked by the conviction that it is not knowledge at all.

It is philosophical analysis itself that has become the principal locus of skepticism. According to the "negative" experimental philosophers, the critical case judgments that distinguish various theories about essential natures—judgments about Gettier cases or preempted causation, for example—are either unreliable or otherwise unfounded (Alexander and Weinberg 2007; Alexander 2012; Stich and Tobia 2016; Machery 2017). The generation of philosophical knowledge is therefore an illusion, and its epistemic pageantry, most notably case certainty, a fraud.

∽

The nativist story says that philosophical inquiry is much like any other sort of inquiry, operating on clues already existing within the mind that have been placed there by some process that is known to be trustworthy. The constructivist story says that philosophical inquiry is quite different from other sorts of inquiry. Rather than investigating the world, it investigates the schemata that we use to organize the world—and to which, as a consequence, we have access that is especially intimate and secure.

This book proposes a middle way. Philosophical inquiry is—at least when it takes the form of analysis—ordinary inquiry, implemented largely in the armchair, into matters of substance. Yet it is poised to succeed, regardless of our starting point, because the philosophical categories are in a certain sense what we make them.

---

[8] There is a sense in which the scientific account is a version of the old story, implying as it does that philosophical knowledge is gained through ordinary thinking—but the package in its entirety, not least in view of its disdain for the armchair, seems too different in flavor from the old story to classify under that heading.

1.6 The Cognitive Way

In the face of confusion, doubt, disagreement, deadlock, the philosopher's characteristic grand maneuver is to step out of the subject matter itself and into "foundations": from mind to metaphysics, from epistemology to logic, from moral philosophy to metaethics. Psychologizing philosophers take a bigger step still, from a subject matter to the science of thought about that subject matter, from thinking about a topic to thinking about the way we think about that topic. This is what I propose to do with philosophical analysis, dissecting our mental representations of philosophical things—our concept of knowledge, our concept of causality, our concept of justice—to see what they are made of, how they are powered, and what trains of thought they draw along behind them.

There is little research on the psychology of philosophical concepts. A great deal of the most influential empirical work on concepts has, by contrast, focused on what psychologists, with some encouragement from philosophers, call natural kind concepts, namely, concepts of species and other biological taxa, such as tiger and swan, along with concepts of chemical substances, such as water and gold.

The psychologizing philosopher ought not to be discouraged, because philosophical concepts and natural kind concepts have much in common. That, at any rate, is this book's unargued working hypothesis: I propose that the conceptual structure, in virtue of which the natural kinds have a palpable out-thereness and substantiality that makes them objects for discovery rather than for stipulation or some other kind of subjective delineation, lends philosophical categories too an objective character that imbues philosophy with its air of inquiry. Philosophizing, then, we feel we are investigating the way things are in the world outside our heads rather than engaging in the sort of free play that is the province of the creative imagination or exercising the clerical punctiliousness with which we might transcribe a mental definition.

The resemblance between philosophical concepts and concepts of natural kinds can be at best partial, however. Inquiry into natural kinds thrives on empirical evidence, whereas philosophical categories—we presuppose and pray—yield their secrets to a kind of ratiocination that can proceed in at least some cases with little or no attention to what goes on in the world of change and decay. (My view is quite different, then, from that of Kornblith [2002], who holds that philosophical categories literally are natural kinds and so should be investigated empirically rather than in the armchair.)

What unites natural kind concepts and philosophical concepts is, in the first instance, their inductivist nature. Inductivism is an empirical thesis about a concept or family of concepts. It consists, as I characterize it, of both a core doctrine and a general attitude toward psychological theorizing about concepts. The attitude is a commitment to explain everyday thinking about a category, as far as is possible, not in terms of the thinker's representation of the category's definition, or in terms of their metalinguistic beliefs about the category's name, or in terms of representations of any other semantic property of the concept, or even in terms of beliefs about the essential nature of the category, but in terms of the thinker's ordinary first-order, non-defining, non-metaphysical beliefs about the category. The core doctrine is the thesis that the beliefs to be avoided—the semantic and metaphysical beliefs—do not, for the most part, exist, or that where they exist they are relegated, like the nobility in a constitutional monarchy, to a peripheral and largely ceremonial cognitive role. At the center of a natural kind concept or a philosophical concept, then, is a set of ordinary beliefs about the corresponding category.

In the second instance, philosophical and natural kind concepts resemble one another with regard to the nature of those central ordinary beliefs: the beliefs concern, in both cases, explanatory connections between category membership and various other properties.

Finally, significant subsets of both the natural kind concepts and the philosophical concepts are similar in that these explanatory relations are secondary rather than primary (a distinction that will be explained in due course). As a result, many—and I would guess most or all—philosophical categories and natural kinds are *secondary categories*, a class of classes that have the Kantian distinction of being both objective and constructed.

All three of these hypotheses—the theses that the natural kind and philosophical concepts are inductive, explanatory, and secondary—are empirical psychological theses (partly psychological in the last case), for which there is in my view considerable if not decisive evidence. My project is predicated, then, on the findings of empirical psychology. Yet it is, at the same time and as my subtitle proclaims, an explanation of the way that even the most empirically indifferent armchair philosophy succeeds in its aims—speaking loosely, an a posteriori vindication of a priori inquiry. If I am correct, then the pursuit of philosophical knowledge can go on as before in its reflective way, largely free of empirical demands. Only metaphilosophers need poke around among the meaty facts in the psychological subcellar.

1.7 The Argument

Here is how the argument unfolds.

In Chapters 2 and 3, I examine two approaches to understanding philosophical analysis as conceptual analysis. What I call the modern approach to conceptual analysis, the epitome of the "new story" sketched above, assumes that our concepts are associated with definitions or other meaning-giving stipulations, such as Kripkean baptisms. (Modern analysis in this latter guise is given the name "intensional analysis.") Hypothetical conceptual analysis, by contrast, assumes that our concepts are associated with nonstipulative beliefs about essential natures. Considerable difficulties that arise in equating philosophical and conceptual analysis are investigated; none, however, turn out to be fatal.

Chapter 4 is a prolegomenon to the use of cognitive psychology for understanding philosophical thought, explaining and to some extent justifying my representationalist and other assumptions. It also comments on the connection between my project and present-day experimental philosophy.

Chapters 5 and 6 turn to the empirical psychology of concepts and discover, in the case of natural kind concepts such as water and swan, that the central assumptions of both varieties of conceptual analysis investigated earlier are false: the natural kind concepts are associated neither with semantic stipulations nor even with beliefs about membership-determining criteria, but only with ordinary theoretical and everyday beliefs. This is the view that I call *inductivism* about the concepts in question.

What if not only our natural kind concepts, but also our concepts of philosophical categories, are inductive? I find this prospect so enticing that I resolve to explore its consequences for the remainder of the book, without stopping to look for further evidence or to argue that it is true.

Given the inductiveness of philosophical concepts, we need a new account of philosophical analysis, I conclude in Chapter 7. In Chapter 8,

1.7. THE ARGUMENT

I provide that account. Analysis is a process of inductive reasoning. We philosophers are like scientific theoreticians who, when all the empirical evidence is in, sit down to inductively infer its implications concerning the structure of the world—except that the evidence is not derived from observation, experiment, or measurement, but from our own case judgments (the so-called intuitions). These case judgments are themselves typically made through inductive reasoning, based on our largely empirically untempered prior beliefs about the philosophical categories in question.

Why trust those beliefs? In answering this question I turn from empirical psychology to philosophical analysis itself. Chapter 9 follows a distinguished philosophical tradition by appealing to the reflexive nature of reference, that is, the fact (so almost any theory of reference supposes) that our beliefs involving a concept play a central role in determining the concept's extension. I propose an approach to reference—the dispositional approach—on which this reflexivity fails to secure the present correctness of our categorizations, but does guarantee their correctness "once all the evidence comes in." I show that this account is able to deal with many problem cases that arise concerning the reference of natural kind terms.

Thanks to reflexivity, we can set out to do philosophical analysis with a clear conscience. But where are we going? Many philosophers are pessimistic about the ultimate prospects for analysis, thinking that it will end not in the proclamation of category-determining essential natures for the philosophical categories but in confusion and despair. Chapters 10 and 11 discuss some reasons for the failure of analysis to realize its ultimate goal, drawn both from experimental philosophy and from more traditional discussions. The last of these reasons is given special attention: many categories resist analysis, Chapter 11 contends, because they have no essential natures to discover.

Without essential natures, what is the point of philosophizing? If philosophical knowledge is not knowledge of natures, what might it be? More generally, what reason is there to think that conclusions arrived at through analysis concern matters of objective substance, rather than reflecting parochial or arbitrary taxonomies projected onto the world by the reflexive aspect of belief?

The purpose of Chapters 12 and 13 is to answer these questions, by sketching a sense in which analysis gives us knowledge of explanatory structure. Essential to this view is the hypothesis that philosophical concepts are not only inductive, but that they are embedded, like natural kind concepts, in explanatory theories. An example—the concept of knowledge itself—is explored.

Chapter 14 confronts the most important difference between natural kinds and philosophical kinds: Why do we investigate the one empirically, the other much less so or not at all? Why is the investigation of the nature of water largely empirically informed, while the empirical contribution to the investigation of knowledge or singular causation is apparently at most a matter of providing necessary scaffolding, such as the validity of belief/desire psychology? I allow that there is not a strict dichotomy to be had. Many philosophical investigations make substantial use of both the observatory and the armchair. But there is a tendency that needs to be explained, one that is manifest in an especially striking way in those many long-standing and central questions in epistemology, metaphysics, and ethics whose answers seem to come to a great extent by way of pure thought.

In the course of this book, I conjecture that many philosophical categories and natural kinds are connected to the rest of the world, according to the theories that constitute our concepts of these categories, by what I call secondary explanatory relations. And I propose, in Chapter 15, that categories picked out by concepts of this sort—the secondary categories—are in a certain sense constructed.[9] They are not a part of, or an aspect of, fundamental reality, but are brought into being by our assembly of the concept. That chapter explains how secondary concepts are acquired and so how secondary categories are ushered into the world, and also how they may be eliminated, if we decide that a category is empty—either empty of members, or like phlogiston nonexistent in some more deeply vacant sense. I remark on the secondary categories' tendency to beget borderline cases, and then provide an explanation of my Chapter 11 thesis that many natural kinds and philosophical categories have no essential natures.

The secondary categories include, I suspect, not only many philosophical categories but also most of the high-level categories that figure in our everyday lives. The theory of secondary categories and their secondary

---

[9] A secondary category is, more exactly, a category that is represented by a secondary concept even when all the evidence is in.

1.7. THE ARGUMENT

reality—given only a preliminary sketch here—is, in that case, a central part of the metaphysics of our human world.

Indeed, although philosophical analysis is my book's official topic, by this final chapter I will have had much to say about representational and epistemic relations between mind and world in general. In that respect, the book can be seen as part of the great modern philosophical project—exemplified by the work of Locke, Hume, Kant, and so many other philosophers—to understand our grasp of the external world by examining the engineering of the mind.

CHAPTER 2

Classical and Modern Conceptual Analysis

2.1 Classical Conceptual Analysis

Twice in recent philosophical history, conceptual analysis has served as philosophy's royal road: in the early modern period, and then again through much of the twentieth century. The first time around it took a psychologizing form that I call classical conceptual analysis, found both in empiricist philosophers such as Locke and in rationalists such as Leibniz and his successor Christian Wolff.[1]

Concepts, held the classical conceptual analysts, are mental representations assembled from more basic representations. Analysis aims to determine the identity of these psychological building blocks, and of the building blocks of which they are in turn composed. A full analysis consists of a complete disassembly of the conceptual "molecule" into unanalyzable conceptual atoms. The end product is not merely a psychological, but also a semantic blueprint: it supplies a definition of the analyzed concept. Why

---

[1] My exposition of classical analysis follows the lines laid down by its foremost enemy, Immanuel Kant; I rely in particular on the interpretations offered by Coffa (1991). A summary of 100 years of philosophy—perhaps *the* 100 years of philosophy—must of course be something of a cartoon. Anderson (2015) gives a sense of the fine historical texture, especially on the rationalist side, and notably places far less emphasis than I do here on the importance of the notion of "thinking in."

should that be? Because when a concept is thought, its parts are "thought in" it; thus, when it is applied to an entity, each of its parts is also applied. If the concept "human" is composed of the concepts "featherless" and "biped," for example, then to think that something is a human is just to think that it is both featherless and a biped. Likewise, to think that some humans have feathers is just to think that some featherless bipeds are feathered—and hence to think a contradiction.[2]

Consequently, it is a conceptual or analytic truth that each of a concept's parts is true of whatever the concept is true of; for example, it is an analytic truth—if the above analysis of "human" is correct—that humans are bipeds. This "conceptual containment" account of analytic truth, further, provides the only possible grounds for such truths, which is to say that all analytic truths are founded in facts about conceptual containment—hence, in psychological facts. The principal aim of philosophy, finally, is none other than the analysis of concepts and the concomitant recovery of analytic truths.

The program of classical conceptual analysis prompts two probing questions. The first, which is the forerunner of the later "paradox of analysis"

---

[2] To a contemporary philosopher, this account of the semantic significance of conceptual containment will seem inadequate in two ways.

First, it appears to imply that a concept's parts are invariably related by simple conjunction. The classical analysts did not, of course, believe this; other relations were obviously possible. But they lacked the semantic sophistication to reconcile their commitments in a single, coherent package.

Second, it supposes that the mind has no choice but to apply a concept's parts whenever the concept is itself applied. Why can't it withhold one of the parts when appropriate (e.g., to wonder whether a human might grow feathers)? To answer in terms of psychological dynamics is, by our modern lights, to run together semantic and psychological necessity. What needs to be added, it seems, is that the parts are essential to the concept: it would not be the concept that it is without them. That requires in turn something to determine in the relevant strong sense "which concept it is"—that is, to determine in effect the concept's individual essence. Such determination cannot be accomplished, on pain of circularity, simply by looking at the concept's parts. It is tempting to get the job done by appending to the concept some sort of explicit stipulation or definition. But of course then containment becomes irrelevant to determining semantic necessity as such (that is, to converting mere ordinary belief, which for a classicist follows from parthood, into something rationally unrevisable).

I suppose that most early moderns would say that concepts are ideas and ideas are individuated by their phenomenology, hence by all their component ideas. But a mere proposal to individuate concepts in a certain way cannot have psychological, let alone semantic, consequences. It is quite coherent for me, as a theoretical psychologist, to count those who believe all swans are white as having a different kind of swan concept from those who believe that some are black. That does not

(Langford 1942), asks why analysis is either effortful or informative. If every time I think "human" I thereby think "featherless biped," why should it require a trained philosopher to recover the fact that the former concept is built from the latter concepts? By Kant's time the widely accepted answer was that the components of our thoughts are not always clear to us, but are rather perceived only "confusedly" (Kant, *Critique of Pure Reason*, B11/A7). What analysis recovers are facts that I know but not transparently:

> If only we knew what we know . . . we would be astonished by the treasures contained in our knowledge.[3]

The second question concerns the relation between the definitions inherent in our concepts and the true natures of the categories to which they refer—the relation between "nominal definitions" and "real definitions." Must the one reflect the other? Perhaps not: perhaps my concept "human" is built from the concepts "featherless" and "biped," yet what in reality makes a thing human—the essential nature of humanness—is to be a rational animal. In such cases, the knowledge available through classical conceptual analysis would be rather shallow, indeed more psychological than philosophical.

In response to this worry, empiricists such as Locke by and large advocated epistemic humility, accepting the limits of analysis. Cartesians relied on God to stock the mind with substantive conceptual structure or the means to attain it. And Kant—Kant destroyed the analytic program, by arguing that the greater and better part of the truths that could be attained by a priori philosophy were not analytic but synthetic, and then showing how a priori synthetic truths could be founded in a constructivist metaphysics of the empirical world.

2.2 Modern Conceptual Analysis

I take up the story again 150 years later, as a new notion of conceptual analysis gathers momentum, reforming the classical notion in the light of

---

prevent the former class of thinkers from wondering whether some swans might after all be black any more than it makes them irrational in doing so.

[3] Kant, *Wiener Logik*, 843; quoted by Coffa (1991, 12).

work by Bolzano and Frege, Moore and Russell, and Carnap and the logical positivists.[4] What I will call modern conceptual analysis is only one among many related ideas about philosophical methodology jockeying for space and time in the middle of the twentieth century, but it commands our attention, first, because it served to some extent as a default view, and second, because it provided exceptionally clear and useful answers to the problems posed in the opening chapter about the status of knowledge produced by analysis—about the reliability of case judgments, case certainty, and substantiality. As such it is as good a starting point for thinking about the nature of philosophical analysis this century as it was in the last.

Modern analysis, like classical analysis, understands the knowledge attained through armchair reflection as founded in conceptual structure. Following Bolzano and Frege, however, concepts are not in the first instance regarded as psychological but as semantic entities: to analyze the concept of, say, knowledge is to determine the structure not of something in the head but of a certain abstract object. The structure of this object, like the psychological containment structure of a classical concept, results in the concept's being associated with a certain definition. That definition dictates the essential nature of the corresponding category and endows various propositions with analytic truth. Philosophical knowledge is knowledge of these definitions or truths.

Because modern analysis opens a gap between what is in the head—the mental representation or "conception"—and what is in the concept itself, there is the prospect of a new epistemological challenge: how can the analyst be sure that their mind accurately represents the semantic structure of the concept itself? Historically, however, from Bolzano until a few decades ago, it has been supposed that there is no real danger of incomplete or faulty representation. The mind contains, it is assumed, a full and accurate depiction of the "objective" semantic structure, along with much that is "subjective" and so should (by the analyst) be ignored.

The significance of the move to semantic space is thus not to create new opportunities for skepticism; it is rather to liberate modern analysis from the containment theory of definitions and analytic truth. Modern analysis, like classical analysis, treats concepts as providing a foundation for analytic

---

[4] As in the previous section, Coffa (1991) is my guide.

2.2. MODERN CONCEPTUAL ANALYSIS

truth by way of definition and supposes that definitions are represented in the mind itself; the difference is that in the modern picture, the definitions need not be reflected in relations of psychological composition. This opens the way to the notion of an "implicit definition," a construct which can bestow a definition upon a psychologically primitive concept, that is, a mental representation that contains no other representations.[5]

I may, for example, introduce a system of days of the week with a set of stipulations: "Monday is followed by Tuesday," "Tuesday is followed by Wednesday," and so on. Collectively, these constitute an implicit definition of what it is for a day to be Monday, Tuesday, and the rest.[6] From the definition certain analytic truths follow: the day seven days after a Monday is also a Monday, and so on. But the stipulation that Monday is followed by Tuesday does not imply that my mental representation of Monday contains the mental representation of Tuesday or vice versa—which is, for obvious reasons, a very good thing.

The modern analyst's final revision to the classical picture is to solve the problem that nominal definitions may not reflect real definitions by way of a thoroughgoing constructivism that in a certain sense simply takes the ordinary understanding of a definition seriously: definitions construct categories by endowing them with an essential nature that is their real definition if anything is. If the concept "human" is built around the definition "featherless biped," then by the force of that very stipulation, to be human is objectively, in reality, to be a featherless biped. It may well be that the members of the category of humans also belong to another category, whose essential nature is to be a rational animal, and it may also be that this latter category is of far greater empirical significance, but nevertheless, to the question "What makes an entity human?" the full and correct answer is "Its

---

[5] A modern analyst is thereby freed, if they wish, to jettison the very idea of conceptual containment; along with it will go any principled distinction between primitive and nonprimitive representations in the psychological sense.

[6] The definition has the defect, note, of failing to fully ground facts about whether a certain day—say, October 24th, 2016—is a Monday. Further conventions are needed to attach the days-of-the-week system to actual temporal intervals (time zones and a dateline will also be useful). You might or might not want to say that these conventions help to define what it is for a day to be Monday, and so to constitute a part of the essential nature of Monday-hood (is 10/24/16 necessarily or only contingently a Monday?); nothing will hang on this.

being a featherless biped." The same is true for the philosophical categories that are the usual targets of analysis: causality, knowledge, the good. The gap between real and nominal definitions is thereby closed.

～

The hypothesis that philosophical analysis is modern conceptual analysis—that it is just what the modern analysts took it to be—has enormous explanatory power.

Before I explore that power, however, I want to make an amendment to the modern picture that will simplify the discussion and foreshadow my own appeal to the empirical psychology of concepts: I want to put aside the notion of concepts as abstract semantic entities and to conceive of modern analysis as built solely on the psychological notion of a concept, that is, solely on certain theses about mental representation.

That may seem like a regressive move. But nothing in the modern picture need be lost. The modern analysts assumed that everything in the abstract concept is also in the psychological concept, so psychological resources ought to be able to replicate the abstract concept's explanatory capacities. The containment theory of definition and analytic truth, meanwhile, can be expunged by banishing the molecular theory according to which concepts are composed of conceptual atoms that are "thought in" every application of the concept itself. As intimated above, we can allow into the mental inventory concepts that have no psychological constituents (or whose constituency is semantically irrelevant) but that ground analytic truths in virtue of their having psychologically real implicit definitions. In this way, we gain all the advantages of the move to semantic concepts without having to posit an additional class of entities.

Of course, semantic concepts may be immensely useful for other purposes, for example, in formal semantics. They are nevertheless not, I think, an essential part of the explanations that matter for the vindication of philosophical analysis, that is, the explanations of the reliability of case judgments, of case certainty, and so on. The wisdom of the retrenchment to a purely psychological understanding of concepts will, I hope, be seen in its fruits—which is to say, in the explanatory enterprises that make up this book.

A classical analyst must deny that every category has a definition: definitions are fixed by containment relations, so there must be some undefined categories corresponding to the conceptual primitives from which everything else is built. A modern analyst, who has no use for containment and who can avail themself of the notion of implicit definition, could in principle hold that every concept is defined (at least, if they are willing to allow that definitions need not provide unambiguous grounding; see note 6). Occasionally in the logical empiricist literature, it can seem as though a philosopher is toying with such a position. (I am thinking of views according to which what counts as observational vocabulary is itself established by convention.) For the most part, however, a suite of semantically primitive concepts is envisaged, typically based in sense experience just as on the early modern empiricist story.

One last supposition: I will assume that when a psychological concept has a definition, it is explicitly represented in the mind (and is explicitly represented *as* a definition). This is not essential to the modern analyst's picture, but it is a rather common assumption that is convenient for expository purposes, as it allows me to delay consideration of the alternatives until near the end of the discussion of conceptual analysis in Chapter 3.

Consider, then, a version of modern analysis that plays out as follows. For each philosophical category that might be subjected to analysis—knowledge, causality, and so on—a psychological concept is posited along with an associated definition, implicit or explicit, explicitly represented in the mind. (That a system of definitions is in the modern analyst's sense "implicit" in no way precludes its being represented explicitly as a set of stipulations, in the mind or elsewhere: "Monday is followed by Tuesday" can be written as "Let the day after Monday be 'Tuesday,'" and so on. Don't blame me; I didn't invent the terminology.) The definition stipulates the criterion for membership of a category—it stipulates what it is for something to belong to the category—and so fixes the category's essential nature. That is the property the analyst seeks to learn. But (says the modern analyst) the definition is "veiled," as in the classical story; it cannot be learned by introspection. Learning therefore proceeds by an inductive process—the method of cases.

It is the familiar story. A scenario is described; a judgment is made as to whether or not a certain specimen in the scenario falls into the category under analysis—whether or not a certain belief is an item of knowledge, or a certain event is a cause—and these judgments (or, more exactly, their contents) are used as evidence for or against various hypotheses about the category's definition, hence its essential nature.

As I have set up the modern ideal, conceptual analysis is a kind of empirical learning that uses psychological evidence—the case judgments—to test psychological hypotheses about the definition-determining elements of psychological concepts. As Frank Jackson writes, it is "hypothetico-deductive" (1998, 36).

From a well-evidenced psychological hypothesis, a metaphysical hypothesis may then be inferred, which is to say, a category's essential nature may be read off from its hypothesized mental definition. (The more traditional conception of modern analysis, involving nonpsychological semantic concepts, can be understood as inserting an additional step: from the putative mental definition, putative semantic facts are inferred, and from these semantic facts, a metaphysical fact.)

At the heart of the theory of modern conceptual analysis is the conjecture that philosophical analysis, at least when it works well, is conceptual analysis: it is an attempt to learn mental definitions and to deduce essential natures therefrom. That conjecture, as we will now see, provides simple and powerful vindications and explanations of many features of philosophical analysis.

Case Judgments Are Reliable. Case judgments are made, a modernist will naturally suppose, by determining whether a specimen fits the corresponding mental definition: Does this belief satisfy the definition of knowledge? Does this event satisfy the definition of singular causation? These definitions are authoritative: they are (by the modern analyst's constructivist lights) real definitions as well as nominal definitions, reflecting and indeed determining the corresponding category's essential nature. The reliability of case judgments, then, depends only on the reliability of our decision procedures for determining the satisfaction of a definition. Such procedures are, at least in the minds of PhD-holders and other leading citizens, presumably highly dependable. Our case judgments will therefore seldom go amiss: "He that Reasoneth aright in words he understandeth, can never conclude an Error" (Hobbes, *Leviathan*, chap. 46, 367).

A modern analyst need not maintain that definitions are directly consulted in every judgment about category membership. Some case judgments are "quick and dirty": I see a canine profile emerge from the blackberry bushes and immediately think "dog." I might retract that judgment on taking a closer look. (Bear cub!) For philosophical purposes, what matters is that these judgments are ultimately supervised by definitions, and that difficult or important cases, in particular, are referred to the final court of definition. The modern analyst cares only, in the end, about scrupulous, deliberate judgments conducted under favorable conditions—conditions that enable what Hobbes described as "reasoning aright."

Case Judgments Are Justified. Insofar as we are justified in believing the modern analyst's psychological theory of case judgments—that they are driven by, or at least supervised by, definitions—we are justified in believing that they are reliable. That is enough, I presume, to justify the judgments themselves, at least among methodologically thoughtful philosophers. (For the others, see below.)

Case Certainty. Why do I feel not only justified in my judgment that a Gettierized belief is not knowledge, but more or less certain? Why do I feel that simply thinking about the case is enough to supply an incontrovertible answer, one that I will not need to revisit—putting aside extreme skeptical scenarios—no matter how much more I learn about the world?

If you are sure that your specimen fits a category's definition, then you can be sure of category membership: the definition's declarations about membership are apodictic, and cannot be refuted or undone. Thus, the modern analyst's definition-based theory of concepts, supplemented by a Hobbesian confidence in our power to determine the satisfaction of definitions, looks to bring such certainty, or near-certainty, well within reach.

The epistemic soft spot in the picture is our knowledge of the properties of the specimen itself, which are often contingently possessed. How can you be so sure that the specimen has the definition-satisfying properties that it seems to have?

As analysis is usually conducted, the facts in question can be known for sure, for one of two reasons. First, the properties of the specimen may be stipulated: rather than putting a real animal or justified true belief in front of you I initiate a thought experiment in which I ask you to imagine and to make case judgments about an organism or belief with certain properties. The only substantive question in such experiments is whether the properties in question imply category membership or otherwise. Second and alternatively, when asking whether, say, stools are chairs, the properties of the class of objects in question are inferred (in the modern analyst's picture) from the class's definition. Everything I need to know about stools to make the case judgment, I can extract from the definition around which my stool concept is built. Thus, I reach the judgment that stools are not chairs by applying deductive logic to the two definitions in question.

Need for the "Method of Cases". When you ask me what makes a number prime, I do not need to run through the paradigm cases—7, 19, 23—asking myself what they do and do not have in common. I simply deliver up to you my mental definition in as many words. Why can I not do the same with philosophical definitions?

The classical conceptual analysts hypothesized a mental veil shrouding the contents of concepts, through which their constituents could be only dimly discerned. It was the philosopher's difficult task to discover the structure hidden by the veil. They did not, in general, proceed by the method of cases, however, but by concerted introspection. Because a concept's components are "thought in" the concept itself—because to think the concept is to think its components—a careful attention to the phenomenology of thought could bring conceptual structure to light.

Modern analysis posits a veil that is not translucent but entirely opaque. Responding to the paradox of analysis and similar concerns, John Rawls (although he was not, as I will later explain, a modern conceptual analyst himself) suggested that the definitions around which our concepts are built are not consciously accessible—they are not available to introspection (Rawls 1999, 41). They are like the principles of syntax in our grasp of natural language: though they play a central role in cognition, we can know them only by their effects. In the linguistic case, the effects in question are judgments about grammaticality. In the philosophical case, they are judgments about category membership. In both cases, the nature of the central principles must be reconstructed from knowledge of the classifications they make about particular specimens. Thus like the study of syntax, conceptual analysis must proceed by way of a hypothesis-testing procedure.

Armchair Availability of Philosophical Knowledge. How can something new be discovered in the armchair? The substantive part of modern conceptual analysis is an investigation of veiled structures in the mind. Because the target of inquiry is in the mind, it is in the armchair, and so can be found there. Because it is, before its recovery, obscured, it may seem as new and surprising as any novel discovery.

2.3 Difficulties with and Defenses of Modern Analysis

The modern conceptual analyst's picture makes such simple sense of the armchair elements of philosophical analysis, yet it is generally thought—or felt—to be outmoded and incorrect. Our present-day skepticism is nurtured by three sources of unease:

1. The modern analyst's picture renders philosophical knowledge insubstantial.
2. Modern analysis does not, on closer examination, so easily explain certain features of philosophical analysis.
3. Modern analysis assumes a theory of concepts, as built around definitions, that is untrue to human psychology.

The last of these will be discussed in the next section, the others here. A fourth reason, gathering momentum as I write, draws on the work of experimental philosophers to cast doubt on the integrity of case judgments. The scope of this concern is far wider than modern analysis, however, and so the issue is postponed to later chapters.

Insubstantiality

The first great objection to taking modern conceptual analysis as the source of philosophical knowledge is that such conclusions as can be reached by modern analysis are not substantial but trivial. Thus, they do not constitute philosophical knowledge in the fullest sense.

Why trivial? Because what is provided by modern analysis is knowledge of how we divide the world into categories, not of how the world ought to be divided or how it is objectively divided. Because it is knowledge of our categorization scheme, not of some subject matter that exists independently of us. Because the propositions learned do not constrain the way things are in the world; they constrain only the way we talk about the way things are.

One response to triviality is joyful acceptance. Yes, insubstantiality is an inevitable corollary of the constructivist move by which modern analysis secures an unbreakable connection between nominal and real definitions. But far from being a liability, this Reichenbachian emptiness solves a great philosophical problem as to the subject matter of the truths of logic, mathematics, and metaphysics. What could they be about, if not the empirical world? They are about nothing at all—not in a pathological way, but in a constructive way, in that they are inevitable concomitants to systems of categorization that help us to efficiently organize our empirical knowledge.

The originators of this approach to the subject matter of the "nonempirical sciences," the Kantians, are more measured in their attitude to the solution. They may insist that such knowledge can concern the way things are in the "world of experience." But still, at some level, they accept insubstantiality—though regretfully, regarding the loss of the possibility of metaphysical knowledge as a painful passage in the journey to intellectual maturity, to the Socratic wisdom that is self-conscious ignorance.

It is also possible, however, to fight back against the claim that conceptual truths, knowable a priori by grasping definitions, cannot be substantive. As several philosophers have remarked in the wake of Kripke's (1980) untangling of the a priori, the necessary, and the analytic, there are clear cases in which a proposition "serves to implicitly define an ingredient term and . . . expresses something factual" (Boghossian 1996, 379). "Factuality" is one step toward substantiality. Can further steps be taken?

Boghossian makes his point using Kripke's example of the "meter stick." The Commissioner of Measurements waves a certain stick—stick S—in the air and defines the standard meter by stipulating that the following sentence is true: "Stick S is one meter long at time t." The stipulation is a definition, and as such can be known to the Commissioner a priori, or at any rate in the armchair. Yet it states a fact about the material world.

A glimmer of light for the conceptual analyst? What Frege said about cognitive significance he might also have said about epistemic significance: modes of presentation make a difference. Suppose that toroidal quantum gravity is the true theory of everything. To discover that toroidal quantum gravity explains everything is a monumental empirical success.

To discover that the theory of everything explains everything is not: being a mere logical truth, it is easy to know but by the same token devoid of scientific importance. Now imagine an enterprising and philosophical scientist who attempts to retain the ease without losing the importance by using the Kripkean meter stick trick. She gives the theory of everything a proper name: "Let 'M-theory' refer to the true theory of everything." Now she knows that M-theory explains everything; further, this is no mere logical truth but a very substantial fact—as substantial as, because identical to, the fact that toroidal quantum gravity explains everything. Yet it is the wrong kind of substantiality; her semantic maneuver does not increase her chances of a Nobel Prize in the slightest.

This demonstrates, I think, the power of the argument from closure stated in Section 1.3. Truths that can be learned from conceptual analysis are "closed"; they are like the stamp that punches out the same shape whatever the material. For precisely the reason that the shape can be known in advance, its contours tell you nothing about the substance from which it is struck.

Philosophical Analysis Is Not about Concepts

The second objection is that modern conceptual analysis—or more exactly, the thesis that philosophical analysis is nothing but modern analysis—fails to explain an important facet of armchair philosophical inquiry. This difficulty has been raised in passing by Kornblith (2002, 1) and recently urged by Williamson (2007, chap. 1). Modern analysis has two kinds of subject matter, psychological and philosophical. The modern analysis of causality, for example, is both an investigation of the concept of causality and also an investigation of causality itself—a relation in the world. But analysis as practiced by philosophers, Williamson writes, does not have this dual character; its topic is philosophical but not psychological, causality but not the concept of causality, or so it seems. So philosophical analysis cannot be modern analysis, or indeed any kind of conceptual analysis.

The most effective response to Williamson's objection comes, I think, in two steps. The first step, already accomplished, is to deploy the Rawlsian veil to hide the contents of conceptual definitions out of mental sight. The second step, which will remain useful long after we have put modern conceptual analysis to bed, is to distinguish two fictional characters, the self-conscious philosophical analyst and the ordinary working philosophical analyst.

The self-conscious analyst is an accomplished metaphilosopher, and understands exactly what they are doing when they sink into the armchair and begin to philosophically analyze a category. If the modern picture is correct, then they understand that they represent the category in question using a concept with an associated definition, that the definition determines the essential nature of the category (as definitions do), and that the definition also regulates their case judgments. They know that they can, as a consequence, rely on their judgments about cases to infer the contents of the definition and so the essential nature, and that provided their reasoning is working according to specification and they think hard, this project must succeed. They are, in short, psychologically and therefore epistemically sophisticated about the process of analysis.

The ordinary philosophical analyst, in contrast, may be entirely uninterested in theories of philosophical methodology; certainly, they know relatively little about them. They simply follow the example of other analysts in making judgments about cases and using those judgments to test theories of the relevant category's essential nature, without asking why their judgments are reliable or how armchair reflection could possibly supply substantive knowledge. Their philosophical reasoning will be as fertile as that of the self-conscious analyst, but because the reasoning contains logical leaps, they are not in a position to explain why.

Suppose that there are few or even no self-conscious analysts about. Then, the proponent of modern analysis may say, only a small minority of philosophers (and conceivably none) will think explicitly about concepts when they perform analysis. Only a small minority, that is, will bring case judgments to bear on hypotheses about essential nature in two explicit steps: from judgments to hypotheses about conceptual definitions, and then from definitions to essential natures. Insofar as the rest—the "working analysts"—succeed, they do so because they take a short cut: their definitionless chain of reasoning is an enthymematic version of the self-conscious analyst's two-step.

The modern analyst can now deflect Williamson's observation that philosophical analysis appears not to concern concepts. It is true, they will say, that ordinary working philosophers, in the course of their labors, do not think about concepts, but that is simply because they are in a hurry and the conceptual definitions are veiled. It does not follow that philosophical analysis does not turn on the existence of mental definitions; indeed, a rational reconstruction of, or explanation of the success of, ordinary analysts' thought will certainly contain hypotheses about concepts and the definitions housed therein. A Williamsonian might reply that even the self-conscious analyst does not think about concepts—but though I think this is correct, as a rhetorical strategy it simply begs the question against the modern analyst, for whom self-consciousness requires precisely such thoughts.

The proposed reply to Williamson has a drawback. The modern analyst argues that philosophers' case judgments are justified by establishing that they are governed by definitions. But this derivation of justification works only for the self-conscious analyst. The working analyst, by hypothesis, has no idea where their case judgments come from. How, then, are they justified in relying on them? How does what they infer from the judgments qualify as knowledge?

One response in defense of modern analysis is to concede that the case judgments are unjustified. That is why we need metaphilosophy, the modern analyst continues in a Cartesian spirit: by understanding the process of analysis, we come for the first time to be truly epistemically entitled to the fruits of armchair philosophy.

Another response is to find an alternative source of justification. Williamson's own work may be useful here. He has argued that we are typically unaware of the basis of the reliability of our thought, yet we are typically justified all the same; why hold philosophical thought to any higher a standard?

But a further more serious problem awaits. In making the suggested reply to Williamson, the modern analyst loses what is perhaps the most compelling part of their story: the explanation of case certainty as a consequence of our seeing that a case judgment follows directly from a definition. Ordinary working analysts experience, as far as I can tell, the feeling of incontrovertibility that accompanies certain case judgments. They are as secure in their classification of a Gettier case as non-knowledge as the self-conscious modern analyst. Yet they, unlike the self-conscious analyst, are not in a position to see that it is derived from a definition. What is the source of their supreme confidence?

The best answer I can offer on the modern analyst's behalf is that the special status of a case judgment based on a definition somehow shines through the Rawlsian veil: the working analyst cannot see the definition itself, but they can see the epistemic glow that it bestows on its consequences. They feel certainty, then, without being able to explain its source.

This explanation gets the job done, but at a cost of a certain amount of ad hockery in the draping of the veil. There is more to come.

2.4 The Case against Definitions

Modern philosophical analysis supposes that for every analyzable category there is a concept that has, at its core, a definition (or a network of interlocking definitions, as with the days of the week). If the modern analyst's picture captures the structure of philosophical analysis, then, our minds had better explicitly represent or otherwise implement definitions for the categories that we undertake to analyze: piety, causality, knowledge, chair. Much empirical evidence suggests, however, that there are few if any definitions in the head. Such evidence comes from three sources. Each has its limitations.

New Theories of Concepts

The first source is work over the last few decades on the nature of concepts, which has persuaded cognitive psychologists that the great majority of concepts have a structure other than that attributed to them by the so-called classical theory of concepts.

The classical theory, which dominated psychological work on concepts from the early part of the twentieth century until the 1970s, is glossed by Margolis and Laurence as follows:

> Most concepts are structured mental representations that encode a set of necessary and sufficient conditions for their application, if possible, in sensory or perceptual terms. (1999, 10)

The word "structured" suggests that the representations have the necessary and sufficient conditions as psychological constituents, as in the early modern theory of concepts that features in classical conceptual analysis. But such an assumption plays little or no role in twentieth-century applications of the classical theory in cognitive psychology, and it does not appear in other characterizations of the classical theory, such as those articulated by Smith and Medin (1981) and Murphy (2002).[7] It is better, I think, to regard the classical theorists in psychology as moving more or less in step (if perhaps unconsciously) with the modern conceptual analysts, by putting aside if not actively renouncing the idea that concepts have an internal structure in favor of the idea that they have definitions of some sort or other.[8]

The requirement that the definitions take the form of lists of necessary and sufficient conditions can be interpreted either restrictively or with varying degrees of liberality. Smith and Medin (1981), who gave the classical theory its name even as they sought to overthrow it, construe it to rule out altogether the possibility of disjunctive definitions. (Never let your enemies define your terms.) Others might more generously allow one of the necessary conditions to contain a disjunction, and of course the more lawyerly will point out that any definition at all technically fits the requirement, if only in virtue of being the sole member of a one-item list. As with many high-level theories in science, we might do better to regard the classical theory not as a sweeping empirical generalization but as a research program, in this case one that prioritized the search for simple conjunctive mental definitions couched whenever feasible in observational terms.

Understood thus, the classical theory was not so much refuted as exhausted: it yielded relatively little in the way of empirically adequate definitions and failed to offer an explanation of some of the most interesting empirical discoveries about categorization. Psychologists, finding the search for mental definitions, or at least for mental lists of necessary and

---

[7] Margolis and Laurence do allow for a view in which "one concept is a structured complex of other concepts just in case it stands in a privileged relation to the other concepts, typically by way of some type of inferential disposition" (1999, 5). Note 13 comments on the dangers of such a move.

[8] A brief overview of the beginnings of the classical view in the twentieth century can be found in Murphy (2002). As Murphy remarks, the first "classicist" was Clark Hull, a behaviorist who surely would have rejected the notion of conceptual containment or internal structure. In that one respect, behaviorism has not withered.

sufficient conditions, to be unrewarding, began to look for other kinds of categorization criteria instead.

Conceptual analysts, it seems, should follow suit. Writing about analysis in ethics, for example, Stich (1993) has suggested that many moral concepts have a prototype structure—that the psychological criterion we use to evaluate the moral goodness of (say) an action takes the form not of the sort of abstract, formal rule familiar from conventional moral theory, but rather one or more "prototypically good" acts or act schemas. The process of categorization—of deciding whether some act falls under the rubric of the good—then involves a comparison between the act to be classified and the prototypes of goodness. Acts that are similar enough to one or more of the goodness prototypes and dissimilar enough to prototypes of moral badness or neutrality are counted as good.

If Stich is right, then a modern conceptual analyst who insists that their account of the good should have a classical look and feel—who insists that it comprise necessary and sufficient conditions at a certain level of abstraction—will never hit upon the correct analysis. Ramsey (1998) generalizes the argument to conceptual analysis across the board.

Prototype theory is, nevertheless, compatible with the existence of definitions of a sort. A prototypical concept of the good might define as good an act that is similar in a certain precisely specified way to one or more paradigms of goodness. Indeed, it seems to me that this is how early prototype theorists understood their idea, writing as they do, in a constructivist vein, that the prototype structure of a concept is imposed on the corresponding category. Modern conceptual analysts might very well (as Stich allows) simply expand their conception of what mental definitions, and thus essential natures, could turn out to be. I will take up this story in Chapter 5.

Psycholinguistic Studies

Fodor et al. (1975, 1980) generate a rich set of psycholinguistic data to make a case, as the title of the latter paper has it, "against definitions."

The assumption underlying this work is that "understanding a sentence requires the recovery of its semantic representation," thus recovery of the definitions of the terms in the sentence (Fodor et al. 1975, 515). Various metrics are used to compare comprehension of sentences containing terms that supposedly have definitions, such as "bachelor" or "kill" (often hypothesized to mean something like "cause to die") with comprehension of sentences that are identical except that the putatively definable term is replaced with an uncontroversially primitive (that is, definitionless) term. The question: is there any sign that in one case but not the other a definition is being unpacked in the course of comprehension? The experiments suggest that no unpacking occurs.[9]

These results are not, I think, entirely incompatible with the classical theory of concepts.[10] They show that what is tokened, when a concept is "thought" or a sentence containing the concept is "understood," has no internal structure. This militates against an early modern view such as Locke's or Kant's in which complex concepts are literally composed of the simpler concepts in terms of which they are defined, so that to perform an operation on a complex concept is of necessity to do something with its conceptual constituents. But it is surely possible that the representation that is tokened when a concept is thought is a proxy with a simple structure—much like a word. By analogy with what computer scientists call "lazy processing," retrieval of the definition is left until it is absolutely necessary, if it ever becomes necessary. Psycholinguistic computation, in that case, may not call on the complex structure, and so the complexity or otherwise of conceptual structure will not reveal itself through measurements of real-time language processing.

In making this proposal, I am in effect distinguishing two kinds of structure attached to a single concept. The first might be called *cognitive structure*; it consists in whatever supervises the deployment of the concept in categorization, other forms of inference, and more generally in cognition. Its nature is what cognitive psychologists argue about when they debate

---

[9] In the most compelling experiments, the putative definitions when unpacked do not merely add words to the sentence but change elements of its syntactic structure, because of the way negatives interact with quantifiers (Fodor et al. 1975) or because of the way that definitions separate subjects from their apparent direct objects (if to kill Charles means to cause Charles to die, for example, then the direct object of "kill" is an event rather than Charles himself [Fodor et al. 1980]).

[10] My reasons are broadly similar to those given by Katz (1977), who relies on a competence/performance distinction: a definition spells out what it is to understand a concept fully, but does not specify an algorithm for attaining that understanding.

theories of concepts. If the classical theory is correct, it takes the form of a definition.

The second kind of structure is what is tokened each time the concept is tokened, that is, each time the thinker has a thought involving the concept. Call it, then, the concept's *token structure*. (For a roughly parallel distinction in the psychology literature, see Rips [1995].)

In the early modern view of concepts, cognitive structure and token structure are the same thing, namely, certain simple concepts arranged in a certain way. These constituents not only determine the concept's conceptual role, but also accompany it in thought wherever it goes: token the concept and you token the constituents. A more computational approach to concepts, however, provides ample scope for the two kinds of structure to come apart. Lazy processing, as noted, allows for a token structure that is simpler than (or simply different from) cognitive structure.

Further, there is no reason why ephemeral or unimportant aspects of a concept might not be tokened when it is thought. In Prinz's "proxytype" theory, for example, concepts seem to have many inessential sensory components that are thought whenever the concept is thought—for the concept of democracy, perhaps "lines at a voting booth or a ballot box" (2002, 180)—but which unlike components of a definition may freely come and go.

In short, a concept with a definition has a complex cognitive structure, but a complex token structure does not of necessity follow. The psycholinguistic arguments are therefore, against the classical theory of concepts, indecisive.

Fodor and his collaborators are alive to the possibility suggested here. Indeed, they suggest that their results are compatible with the existence of conceptual necessities that are due to "meaning postulates," which are implicit definitions (more exactly, definition parts) such as "Monday is followed by Tuesday."[11] As far as I can see, a suitable set of meaning postulates could constitute a definition for the corresponding concept. Thus, in spite of their titles, Fodor and his collaborators are not arguing against the thesis that concepts have definitions. They are rather—a closer reading of their opening and closing sections shows—arguing against the

---

[11] The term was introduced by Carnap (1952).

thesis that concepts have definitions in virtue of their "internal structure." This sounds much like the early modern version of classicism, or perhaps the version articulated by Margolis and Laurence, which attributes to concepts not only definitions but "structure."

Given the context, it is perhaps best to understand Fodor et al. as making a case against the 1970s theory of generative semantics, which they take to attribute a structure to lexical concepts that not only spawns definitions but also gets processed in the course of normal sentence comprehension.[12] The classical theory of concepts, at least as understood by cognitive psychologists and as presupposed by modern conceptual analysts—who ought to be as happy to retrieve definitions established by meaning postulates as those established by internal structure—apparently survives.[13]

The Paucity of Analytic Truths

If many concepts have definitions, then there will be many conceptual or analytic truths. The definition itself is one such, of course, but its logical consequences also qualify. If the concept of a human has at its stipulative core "rational animal," for example, then it is a conceptual truth that humans are animals, and likewise that they are rational.

Reports from the field, however—from thousands of analytic philosophers engaged in an unflagging pursuit of the a priori—suggest that there are very few analytic truths to be found. The principal source of evidence: almost no general principle outside the realm of mathematics and logic exudes the aura of infallibility that is guaranteed by analyticity.

Even in the apparently most straightforward cases, exceptions to "analyticities" have arisen. Is every unmarried male a bachelor? Not children; not members of religious orders who have vowed celibacy; not those living

---

[12] I suggest this as a kind of rational reconstruction, since Fodor et al. evidently take their target to be rather broader than that.

[13] You might wonder whether we couldn't define "internal structure" with reference to conceptual necessity: if it is conceptually necessary that swans are birds (because of a definition, a meaning postulate, the conditions for the possession of the swan concept, or whatever), then the bird concept is part of the internal structure of the swan concept (see note 7). Such a maneuver will have many ugly consequences. In the case of days of the week, the concept of Monday will end up containing itself. More generally, motley logical consequences of a concept's meaning postulates will end up "inside" the concept; for example, "bird or football" will end up inside "swan."

in marriage-like arrangements. Nor is marriage an insurmountable barrier to bachelorhood: a man who as a stunt married in Las Vegas one long-ago weekend, and who has gone on to live the bachelor lifestyle without bothering to obtain a divorce, seems to qualify.[14] More philosophically pertinent is the case of knowledge. Is it analytic that a piece of knowledge is true, that it is justified, that it is believed? Philosophers have intelligibly argued against each of these propositions, suggesting that even if correct, they cannot be conceptual truths (Ichikawa and Steup 2014).

If there are few analytic truths, then few concepts can have definitions—in which case modern conceptual analysis is in all but a handful of cases (such as the category of prime numbers) impossible. Harman (1994) makes the argument memorably; others such as Fodor (1998) have long advanced the related argument against the classical theory of concepts.

In parallel to this empirical argument against the prevalence of analyticity, a philosophical campaign against the very possibility of analyticity has been waged. The leader is of course Quine (1936, 1951, 1963). If Quine's argument—ultimately based on rule-following considerations of the sort later promoted by Kripke (1982)—succeeds, then there can be no mental definitions; indeed, neither human minds nor any other naturalistic systems are capable of creating a genuine definition, however hard they try. (Even mathematical "definitions" will turn out to inhabit the gray Quinean land between the unreachable poles of definition and pure empirical proposition.)

Boghossian (1996) has pushed back, seeking to find a place for both implicit definitions (for logic and basic mathematical notions at a minimum) and, at least in principle, explicit definitions, in human thought. (He has also made an influential distinction between "metaphysical" and "epistemological" conceptions of analyticity, which I need not explore here, since I have no use in this book for either notion.) Williamson (2007) has used some of the same tools, however, to argue against the possibility or at least the prevalence of analytic truth.

The result is an uneasy standoff. Where philosophers once confidently cited Quine as the slayer of analytic and conceptual truth, they are now less certain; Williamson, in his own debunking of analyticity, writes that

---

[14] Not autobiographical.

Quine's critique "no longer seems compelling" (2007, 51). Yet skepticism about philosophically interesting analyticities and definitions remains widespread, suggesting that Quine's argument was largely a theoretical cover for a conclusion drawn inductively: the program of looking for analyticities and definitions had too often returned with flawed goods or none at all.

What is the modern conceptual analyst to do? I see two possible strategies.

One move begins with an appeal to the distinction, which I earlier laid aside, between psychological and semantic concepts. Most semantic concepts (the story goes) are or are built around definitions. But psychological concepts often fail to capture the full content or stipulative status of these definitions. The psychological bachelor concept, for example, though attached to the semantic concept (which is what makes it the psychological concept of bachelor rather than something else), does not fully recapitulate its contents in the head. Thus the head does not have complete access to the definition that sits at the conceptual core, and therefore cannot recognize the analytic truths generated by that definition, or at least cannot recognize their analyticity.

Such a maneuver threatens, however, to undermine the entire enterprise of conceptual analysis. On the one hand, it creates even greater difficulties for the modern analyst's explanation of case certainty than were created by the posit of the Rawlsian veil. I return to those difficulties immediately below, so let me not dwell on this problem.

On the other hand, the maneuver (as remarked above) opens up modern conceptual analysis to a skeptical challenge: what if the contents of the psychological concept are not rich enough, or are not accurate enough, to underwrite the case judgments needed for successful analysis? Many decisive moves in analysis depend on judgments about "edge cases" such as Gettierized justified true belief or preemptive singular causes. If the psychological concept, which is all the armchair philosopher has to go on, is only an imperfect representation of the definition, might they not make imperfect judgments about these crucial cases—even if their judgments about ordinary cases are reliable? Then either the psychological concept will not succeed in attaching itself to a determinate semantic concept after all, or even if it is successful, we will have no prospect of determining which semantic concept it has lassoed. An argument along these lines is forcefully pressed against conceptual analysis by Johnston and Leslie (2012).[15]

The second way for the modern analyst to explain the apparent paucity of analyticities is to allow that definitions are spelled out in the head in their entirety while executing a more extreme version of the Rawlsian maneuver suggested in Section 2.2 above: just as we are cut off from direct knowledge of the definitions in our conceptual cores, the modern analyst might propose, we are cut off from the infallibility of the truths that they imply. A consequence of a definition might seem highly plausible to us, but—unable as we are to locate the source of this plausibility in the stipulative character of the definition—we may not feel confident in declaring it to be immune from all rationally possible doubt.

The more lavish the modern analyst's use of the Rawlsian veil, however, the more difficult it is to explain case certainty, that is, to explain our sense that judgments about some examples, such as standard Gettier cases, cannot be mistaken.

The original modernist account of case certainty was straightforward. Our case judgments are applications of definitions, according to that picture, and definitions are beyond doubt, so in cases where we have enough information about a specimen to see that the definition either certainly is or certainly is not satisfied, we can have no doubt about the ensuing judgments. Two assumptions are made in the course of the explanation: that the categorizer knows the contents of the definition, so that they can distinguish judgments that follow deductively from the definition from those that follow only with probability, and that they know it is a definition, so that they know what follows deductively follows with certainty.

Even the original Rawlsian veil obstructs this story—the veil's dialectical function, after all, is precisely to obscure from us the fact that a proposition with a certain content is category-defining. Some sort of repair job is needed. I suggested above, on the modern analysts' behalf, that our mind

---

[15] Johnston and Leslie assume, on behalf of the conceptual analyst, that semantic concepts are built around "application conditions" that deliver apodictic judgments about category membership. (These might be definitions or they might be "intensions" of some other sort.) They go on to provide what they take to be empirical evidence that psychological concepts do not entirely capture the content of anything like a set of application conditions; the skeptical argument proceeds from there.

does some unconscious epistemology, transmitting to a definition's logical consequences a sense of immediate certainty that is palpable even though the grounding definition itself is not clearly grasped. Thus we (correctly) feel certain of the consequences of a definition without seeing clearly the epistemic grounds of our certainty. When a consequence is a case judgment, this absolute confidence amounts to case certainty.

I have now suggested a more liberal deployment of the Rawlsian veil to explain how there can seem to be so few analyticities given that there are so many definitions: the veil hides the infallibility of a definition's logical consequences. This clearly upends the revised explanation of case certainty.

In a last-ditch defense, a modern analyst might attempt to rearrange the veil so that it disguises the infallibility of general principles but not of judgments about individual cases—an ad hoc cover-up worthy of a cognitive Christo.

Or they might conjecture that we simply assume that our case judgments, but not our more general beliefs about a category, are based on definitions—albeit unknown—and so are infallible. This suggestion implies, rather implausibly, that we are all tacitly modern conceptual analysts and that our sense of certainty in making a Gettier judgment is based on a sophisticated but apparently unconscious piece of psychologically inflected metaphilosophy.

The rarity of analyticities may not constitute a knockdown argument against the existence of definitions, then, but it compels the modern analyst to squirm in an unseemly way. Can some alternative to the modern analyst's picture save the dignity of conceptual analysis? Read on.

CHAPTER 3

Other Forms of Conceptual Analysis

3.1 Intensional Analysis

If concepts do not have definitions, how can we be so sure of our case judgments in philosophical thought experiments—how can case certainty be explained? Such certainty implies, it seems, that we have in our heads a criterion for category membership that cannot go wrong, and that we know cannot go wrong. It need not be a definition.

Consider, for example, a simple and highly idealized version of the causal-historical theory of reference for natural kind terms. Having encountered various specimens of a shiny yellow metal, I say to myself: I am going to use the word "gold" to refer to anything having roughly the same microstructure as these specimens (gesturing to the specimens). From this referential intention, "gold" in my mouth comes to refer by fiat to a certain substance. The intention might not, however, fix a definition for "gold"; that term might for some reason have a Millian semantics.

Suppose that I have in my head a concept corresponding to the word "gold" and having the same semantic properties. This gold concept, being Millian, has no definition. But I do represent to myself a rule that, by stipulation, determines its extension, namely, the rule spelled out by my referential intention. Understanding this intention, and grasping its stipulative status as a reference-fixer, I can come to have armchair knowledge that a certain criterion determines whether or not something is gold; namely, its having the same microstructure as the ostended specimens. This criterion therefore spells out gold's essential nature—although learning enough about the world to see exactly what microstructure is picked out is, of course, a difficult empirical endeavor.

I am tempted to call these stipulative reference-fixers *intensions*. I would then be using the term "intension" in its old-fashioned sense, as something that determines an intension in the more modern sense—that is, something that determines a mapping from sets of information or scenarios or worlds to extensions. To avoid confusion, I will resist the temptation. But I will allude to it by giving the name *intensional analysis* to a certain theory of philosophical analysis.

Intensional analysis in its simplest form is exemplified by the case of "gold" as told above. For each category to be analyzed, it is supposed, there is an explicit stipulative reference-fixer, that is, a rule in the head that spells out a criterion determining the corresponding term's extension. Such reference-fixers exist, in particular, for philosophical terms, and they directly determine, in their constitutive way, the philosophical categories' essential natures.

Intensional analysis of a category attempts to recover the reference-fixing rule and to read off from the content of the rule the category's nature. It cannot do so by direct introspection: reference-fixers, like the modern analyst's definitions, are veiled (so goes the story). It therefore uses case judgments to infer what's in the rule.

The steps from judgments to rule, and then from rule to essential nature, are undertaken with complete confidence (or close enough), the first because the case judgments are through some psychological mechanism or other produced in conformance with the rule, and the second because the concept's intensional rule by its stipulative role cannot but articulate the category's nature, understood as the property in virtue of which specimens fall into the category.

Pretty much everything that the modern analyst gets from conceptual definitions, the intensional analyst gets without them. The theory of intensional analysis must perform a similar balancing act, however, between on the one hand the psychological transparency required for armchair philosophy to retain its claim to a priori knowledge, and on the other hand the

Rawlsian veils required to explain why analysis must go by way of cases and why there are apparently so few analytic truths—for there is no reason in general to think that stipulative reference-fixers will spawn fewer analyticities than do definitions.

3.2 Stipulativity

As I have characterized them so far, both the modern and the intensional interpretations of conceptual analysis turn on four theses:

1. The mind explicitly represents the term in question (that is, the term that picks out the category under analysis) as having a certain semantic feature, either a definition or a reference-fixing rule.
2. This representation has an explicitly stipulative nature, as a consequence of which it does not merely attribute, but successfully attaches, the semantic feature to the term. It not only says that the term has (say) a certain definition; it makes it so.
3. The content of the representation plays an authoritative role in generating or at least in supervising case judgments.
4. The stipulative nature of the representation is sufficiently transparent to the user to justify case judgments and to explain case certainty.

Analysis then proceeds in the obvious way. First, the content of the semantic attribution is learned, because of the veil either in the main or entirely through consideration of case judgments. Second, through some process or other the stipulative status of the attribution is grasped. Third, from the constructive power of the stipulation the analyst infers that the corresponding category necessarily possesses certain properties. Whether what is stipulated is a definition or a reference-fixing rule, the properties in question presumably constitute the category's essential nature: they are the properties, possession of which makes something a category member.

In order for analysis of this sort to be possible, must the property attribution assumed in (1) be explicit? Must the stipulative status assumed in (2) be explicit? No to both questions, as I will explain in this section with the help of several examples; both the attribution and the stipulative status may be tacit. (I avoid the otherwise preferable "implicit" to avoid confusion with implicit definitions.)

The early modern molecular theory of concepts provides a first example. According to that theory, a concept behaves in thought as though it is defined in terms of its constituents simply because the constituents are "thought in" every application of the concepts. The concept of gold, for example, might be made up of the concepts yellow, metallic, and malleable; as a consequence, to think that a specimen is gold simply is to think that it has those three properties (Section 2.1). The simple constituents of a complex concept thereby function as a definition of the concept. But there is no explicit representation of the definition as such. A molecular thinker, unless they are philosophizing, does not think to themself "I define 'gold' as follows. . . . " They do explicitly attribute the properties to gold (believing, for example, that all gold is yellow), but this judgment does not differ psychologically from nonstipulative attributions such as the belief that all gold is valuable. It takes a certain psychological sophistication—indeed, a certain theory of concepts—to understand that, given the way the mind works, it is logically incoherent to think that some gold might not be yellow, since that would be to think that some yellow, malleable metal is not yellow. (Perhaps not too much sophistication is required, if the structure of such thoughts is readily apparent to any attentive thinker.)

A molecular thinker, then, reasons *as if* they had stipulated to themself that (say) "gold" is defined to be yellow, and they are capable of coming to see that their mental economy is set up so as to have this consequence. Even if there is no act of stipulation, then, they can engage in philosophical analysis more or less as the modern or intensional conceptual analyst conceives it: they can come to know through introspection or reflection that their "gold" has certain semantic properties in virtue of its cognitive role, and from this more or less certain knowledge of the semantics of "gold" they can infer certain necessary truths about gold itself, most importantly its essential nature.[1]

In such cases I say that a semantic property of a term, such as a definition or intension, is *stipulative*, or possessed *stipulatively*, even in the absence of

---

[1] The molecular theorist's inference from constituency to necessity is not, I think, airtight (see note 2 of Chapter 2), but it is related here as an instructive attempt.

an actual stipulation or representation of stipulative status. This "stipulativity" means, then, that the term has the property and that its having it is in some sense transparent to the user, or at least that the user reasons as though it were transparent. The "in some sense" and "as though" allow a certain latitude for exploring various forms of stipulativity and their usefulness in explaining philosophical analysis. One important question, for example, is the degree to which the "as though" confers or otherwise comes with justification.

A second example of tacit "stipulativity," to be considered in greater detail later, can be drawn from attempts to provide an "inferentialist" theory of meaning or concept possession. As a very simple example, consider the following theory. To grasp a concept is, constitutively, to be disposed to make certain inferences concerning the corresponding category. In some cases, these inferences might simply be, to use Peacocke's (1992) term, "primitively compelling": they are not made on the grounds of a definition or any other suppositions or beliefs. Nevertheless, they might be precisely the inferences that are warranted by a certain definition together with the laws of deductive logic. The possessor of the concept is therefore psychologically disposed to reason just as though they represented a definition to themselves, and—because it is precisely that disposition that is constitutive of possessing the concept—they are in fact justified in reasoning in this way (so the argument goes, loosely paralleling the argument made by the molecular theorist). Thus, the definition, though neither its content nor its stipulative status is explicitly represented, satisfies the conditions for stipulativity.

In the molecular and inferentialist accounts, concepts acquire their stipulative semantics—their definitions or intensions—in virtue of their intrinsic properties, in the one case by their literal psychological contents and in the other by primitive inferential dispositions. But stipulative status might also be attained as a result of interactions between a concept and other concepts, beliefs, and further aspects of the machinery of thought.

One such source of stipulativity is commitments about the nature of reference. To present the idea, let me consider an extremely crude proposal; some sophistications will be discussed in Chapter 7.

Suppose that there is no interesting intrinsic difference between your belief that gold is malleable and your belief that gold is valuable. Both have the same "mental syntax" and feature in reasoning in similar ways; there is no sign, in particular, that the concept of malleability is constitutive of the gold concept in a way that the concept of value is not, nor are there any primitive differences in inferential valence between the two. The former might nevertheless acquire a special semantic significance if you make the following stipulation to yourself about the way that your chemical substance terms (including "gold") refer: you declare that these terms pick out all and only those things satisfying your beliefs about the corresponding category's chemical properties. Malleability is a chemical property but being valuable is not. Thus your malleability belief plays a role in determining the reference of "gold" that your value belief does not; as a consequence, the reference of "gold" is fixed so that gold is necessarily malleable, but not necessarily valuable.

Although the malleability belief does not itself bear any markers of stipulativity, in virtue of a separate stipulation about the nature of reference it nevertheless confers on the term "gold" a semantic property with the force of a stipulation. More generally, your chemical beliefs about gold in effect indirectly stipulate, by way of your prescription about reference, a certain reference-fixing rule for "gold," thus determining for it a corresponding intension.

This picture yields a particular route to understanding philosophical analysis as intensional analysis: knowledge of the nature of reference, and of that nature's stipulative status in particular, gives you access to both the content and the stipulative status of the reference-fixing rules for a wide array of concepts, including philosophical concepts, and so makes intensional analysis possible. This I call the "reference-first" conception of intensional analysis.

The reference-first conception and other indirect conceptions of stipulativity suggest an approach to explaining the Rawlsian veil. In the reference-first story above, my mental representations collectively determine that "gold" refers to (let's say) yellow, malleable metals. But without reflecting on reference, I may not notice this consequence. My stipulative reference-fixer for "gold" is blending in with the crowd, looking for all the world like an ordinary, nonstipulative belief to the effect that gold is yellow, malleable, and

3.2. STIPULATIVITY

metallic. What is veiled is its stipulative status, which is fixed by an entirely separate belief that does not even concern gold as such.

The veil is hazier still if what is indirectly stipulated to determine reference is not spelled out in a few simple beliefs. Suppose, for example, that you stipulate to yourself that your words refer to whatever you would apply them to, were you to come to know all the fundamental physical facts about the world.[2] For a kind term such as "gold," the resulting reference-fixing disposition—the psychological fact as to how you would apply "gold" were you to learn all the fundamental physical facts—might be determined by complex interactions among your current beliefs, inferential style, and values. Perhaps the only practical way to investigate its nature is to put hypothetical cases to yourself, asking yourself what you would count as gold if the world turned out to be such and such a way—just as in philosophical analysis. Complexity would in that case provide the conceptual analyst's Rawlsian veil.[3]

There is a weakness, however, to these attempts—interesting though they are—to account for the Rawlsian veil: though they are consistent with case certainty, they make it too difficult to achieve. The self-conscious analyst is well positioned to attain case certainty: they perceive the stipulative nature of the criteria that determine their case judgments, and conclude that those judgments cannot err (provided that their information about the specimen in question is correct). But the ordinary working analyst, who by assumption has not reflected on the veiled source of their judgments, cannot reach certainty by this route. Where, then, does their confidence in their "intuitions" about category membership come from? How can they be so sure that their judgments are not based on a criterion for category membership that is, even if generally reliable, nevertheless fallible?

---

[2] A more sophisticated view along these lines would turn on your dispositions were you to learn not only the basic physical facts but certain others, as explained in note 2 of Chapter 7. An alternative dispositional account of reference that turns on observable rather than fundamental facts will be proposed in Chapter 9.

[3] Chalmers (2002, 148) goes so far as to suggest that some such dispositions might not be able to be "captured by a description." It would follow that a complete human psychology cannot be captured by a description, something that I find difficult to countenance. But it is possible, given the context, that Chalmers means only to say that the effect of the dispositions might not be captured by a relatively short description.

One way out of the quandary is to adopt the full inferentialist package and to build certainty into the inferential dispositions that result in philosophical case judgments, and ultimately into primitive inferential dispositions. Part of having the concept of knowledge, then, is to feel certain about judgments about Gettier cases. I will not try to undermine this move here, but I do note that it results in a rather disappointing psychological explanation of certainty, and one that is unable to say anything interesting about variations in certainty for different, related scenarios (for which see Chapter 10).

Another move, which is more useful to consider at this stage, is to deny there is anything to explain: case certainty as I have described it is confined, at best, to metaphilosophers and other self-conscious analysts. What other analysts have is not case certainty but a kind of default case confidence. They regard their judgment that a standard Gettierized belief is not knowledge just as they treat most of their perceptual judgments: provided that no awkward questions arise and that conditions seem to be generally favorable, they have no reason to question these judgments, and so they do not. There is no commitment to the irrelevance of further information, but in the absence of such information, the judgments have, like many non-apodictic judgments, a settled feel. And that is all there is to "case certainty" for the working analyst. It is this possibility that I would now like to explore.

3.3 Hypothetical Analysis

Is there a way of understanding philosophical analysis as conceptual analysis—as a technique for discovering philosophical knowledge that goes by way of knowledge about concepts—that does not elevate the conceptual content in question to apodictic status? Is there a kind of conceptual analysis, that is, which does not suppose that conceptual content supplies infallible rules for making case judgments or imply that it entails myriads of conceptual truths? There is; I call it hypothetical conceptual analysis, or for short, *hypothetical analysis*.

Like modern conceptual analysis, hypothetical analysis assumes that we represent essential natures. Within each philosophical concept, then, lies a statement of the corresponding category's essential nature. Whereas modern conceptual analysis confers on this statement stipulative status and thus infallibility, hypothetical analysis presumes that it has, in many cases, no more than suppositional status. It is a hypothesis about the category's nature—a metaphysical posit, but one that is subject to correction by further reflection or other investigation. Perhaps such posits, like the modern analyst's definitions, never in fact change; perhaps there are no likely refuters or underminers on the horizon. What distinguishes hypothetical analysis is that they are not rationally immune to change, or more exactly, to refutation. They have the character of beliefs, not pronouncements.

Hypothetical analysis assumes, then, that for each philosophical category that is susceptible to analysis, we have a non-apodictic representation of the category's nature. Call this the corresponding concept's *essence postulate*. The aim of analysis is to reveal the contents of the essence postulate. Because it is hidden behind one of those Rawlsian veils, the analyst has no direct access to the principle. But they have indirect access, because it is the essence postulate—the believer's representation of the category's nature—that guides categorization, determining which specimens are judged to fall into the category. (As with stipulative principles of categorization, this guidance may very well be hands-off: the role of the essence postulate may be to calibrate the heuristics that are used in day-to-day judgments of category membership.) The nature of the essence postulate can be discovered, then, by using case judgments as evidence to sort through various hypotheses as to the postulate's content. Conceptual analysis of the hypothetical variety is precisely this endeavor.

Are there self-conscious hypothetical analysts? John Rawls, I think, is one. I have in mind his insistence that the philosophy of justice turn not only on definitions but on our preexisting commitment to substantive moral principles (Rawls 1999, 44–45), represented in the mind but not directly accessible to introspection. Rawls's particular strategy for getting at the hidden but substantive heart of a concept is what he calls reflective equilibrium; I find in his description of that method something not too different from a sophisticated version of the process described in the previous paragraph, in which case judgments about particular specimens serve as the "data" supporting or undermining hypotheses about the contents of essence postulates. The sophistications are familiar from scientific inquiry: data, hence case judgments, may be rejected if they are either explained away or if their evidential significance is outweighed by global considerations of simplicity and unity.[4]

To the question, then, whether philosophical analysis can be interpreted as some form of hypothetical analysis. By abandoning stipulativity, the hypothetical analyst destroys the strong link between nominal and real definitions, that is, between the criteria for category membership represented by a concept, on the one hand, and on the other, the nature of the category picked out by the concept. If an essence postulate is a mere opinion about a category's essential nature, might it not be a radically false opinion? In that case, even if hypothetical analysis produced reliable psychological information about the contents of concepts, it would fail to produce reliable information about the nature of the corresponding categories. So long as this remains a real possibility, beliefs arrived at through hypothetical analysis cannot be considered to be justified. They might be correct, but they do not amount to philosophical knowledge.

A parallel problem arises for explaining our confidence in case judgments. It is a characteristic of hypothetical analysis that it does not understand case judgments to be infallible—based as they are on an essence postulate that is itself corrigible. That will be a fatal concession, however, if in relinquishing palpable infallibility hypothetical analysis also relinquishes palpable reliability, that is, if in renouncing the claim to certainty about case judgments it thereby loses its claim to confidence of any sort whatsoever.

These are, in effect, two consequences of a single difficulty, that of finding a reliable link between essence postulates and essential natures. Perhaps we simply tend to have high confidence in our essence postulates? That would explain our confidence in both our case judgments and in the philosophical theories we infer from the essence postulates suggested by those judgments. But to attain philosophical knowledge, rather than mere philosophical opinion, we need some justification for this confidence, some reason to suppose that our essence postulates are typically on target.

---

[4] Williamson (2007, §7.7) is surely correct that the usual characterization of reflective equilibrium is loose enough to fit many rather different conceptions of philosophical inquiry; the strategy I attribute to Rawls is just one such method. Another is Weatherson's (2003) proposal, to be sketched shortly, on which the balancing of intuitions about particular cases against broader theoretical concerns has nothing to do with empirical inquiry into conceptual contents but is rather a purely deductive undertaking, namely, an attempt to apply to a concept a reference-fixing schema that calls explicitly for such a balance.

It is possible to give a nativist justification, arguing that the accuracy of essence postulates is guaranteed, or at least made probable, by the processes that inveigled them into the mind—evolutionary, demiurgical, or whatever. But more in tune with the constructivist story about philosophical knowledge offered by modern conceptual analysis is a rationale of the following sort: a concept's essence postulates are very likely accurate because they play a central role in determining the essential nature that they purport to articulate.

A hypothetical analyst taking the constructivist line cannot of course hold that a concept's essence postulate by definition dictates the essential nature of the corresponding category. That is the modern analyst's territory, with all its pitfalls and boobytraps. What they might say instead is that the essence postulate plays an important, but not incontrovertible, role in determining the extension (or intension) of the concept, and so the nature of the category. Suppose, for example (loosely following Weatherson [2003]), that reference works like this: a concept's corresponding category is determined by a weighted mix of essence postulate, other beliefs, and "naturalness," with the essence postulate counting for most but not all of the mix. Then you would have some reason to think that your essence postulates tended in most cases to capture the nature of the categories about which they make their claims. Hypothetical analysis responsibly pursued would therefore offer philosophical as well as psychological knowledge.

Such a strategy for vindicating philosophical analysis has two problems. (A third problem, that of the substantiality of the knowledge gained, will be raised in connection with my own related view in later chapters.) First, it suggests that philosophical analysis, if it is to move securely from knowledge of the essence postulate to knowledge of the essential nature itself, should involve reflection about the nature of reference, and self-conscious application of the results of this reflection. But many analysts would deny that the philosophical conclusions they draw are made with the help of a premise about reference, just as they might deny that their procedure involves reflection on the structure of their concepts or the content of their beliefs. This latter problem was solved in Section 2.3 with the help of a strategically deployed Rawlsian veil; it looks to be much harder, however, to conceal the potentially embarrassing apparatus in the present case, that is, to hang a veil in such a way that the mind contemplates what mix of essence postulate and other factors maps, according to the prevailing reference schema, to what category, without being in the least aware that it is thinking about mixing, weights, or reference at all.

Second, if in the formula for determining what essential nature is picked out by what concept, the essence postulate is weighted heavily enough that we can in general trust the results of our analyses, then it seems that we cannot, except in very unusual circumstances, be seriously mistaken about the essential nature of the categories to which our concepts refer. But it seems that we often do make such mistakes. The spectrum of opinion among philosophers as to the essential nature of most philosophically interesting categories is so wide that a good majority of philosophers must be wrong most of the time. On the hypothetical analyst's story it is hard to see how this is possible: given the strong constructivism implied by the heavy weighting of the essence postulate, it must surely be the case that two philosophers in deep disagreement are most likely talking past one another, using the same word to refer to two different categories, each answering dutifully to its own proprietary essence postulate.

An unpleasant prospect, I think; if there is something much worse than a lack of consensus, it is mutual unintelligibility. To preserve the possibility of philosophical debate that is deep and real, the hypothetical analyst must, it seems, weigh essence postulates more lightly—but then the connection between conceptual analysis and philosophical truth is quite possibly lost.

༄

In the next four chapters, I will take a different tack against hypothetical analysis, and at the same time against modern conceptual analysis, intensional analysis, and any other view in which philosophical analysis involves the analysis of concepts.

Most or all of our concepts of categories, I will argue, including those of the philosophically interesting categories, contain no essence postulates, no definitions, and no reference-fixers, either explicitly or tacitly stipulative. Our determinations of category membership—our case judgments—are not based on some belief or stipulation about the essential nature of the category, or about its referential connections to the world or other semantic properties; nor are they powered by a psychological disposition with defining or reference-fixing aspect or status. No such things inhabit the heads of ordinary reasoners, and those found in the heads of extraordinary reasoners play a far less important role in categorization than their supreme logical station would suggest.

It is therefore impossible to read putative essential natures for categories off of our conceptual representations. Were modern analysis, intensional analysis, or hypothetical analysis to be implemented self-consciously, they would fail completely. Philosophical analysis is, by contrast, far from being a complete catastrophe. Whereas conceptual analysis must return empty-handed, philosophical analysis brings back many valuables, if not always the grand metaphysical prizes for which we hoped. Consequently, there must be more to philosophical analysis than conceptual analysis.

CHAPTER 4

The Psychology of Philosophy

On the first day of creation what philosophers do is to philosophize—often blithely, guilelessly, without self-doubt or second thoughts. Pause to think about the process, however, and the urge mounts to psychologize, that is, to ask: What is going on in the head when we make case judgments or use those judgments to formulate hypotheses about essential natures? What inferences are we making, in accordance with what principles, from what premises, and using what concepts? Which elements of those inferences are known to us, and which are hidden by multifariously draped "Rawlsian veils" and the like?

We should just eat the apple—that is, we should succumb to the temptation to psychologize. It is not only delicious but good for our methodology. That is the leading premise of this book. By attempting a science of philosophical thought, we can better understand our thinking not only from a psychological but from a philosophical perspective, addressing questions about accuracy and justification, truth and substance, that are the proper concern of the philosophy of philosophy.

Some of the assumptions of my psychology of philosophy are laid out in this chapter. They are for the most part working hypotheses, stated but only minimally supported. That is in accordance with the "what-if" tenor of the project: my aim is to explore new and promising ideas about the way that the philosophical mind works, inspired by but not limited to what the empirical data concerning the nature of concepts and inferences has so far shown.

4.1 Concepts

Concepts are what we think with; to think about swans, say, I use my swan concept. Such a platitude is consistent with two different ways of understanding the mental role of concepts.

On one approach, my swan concept is whatever I token when I think about swans. It is like a mental "word," a constituent of more complex, proposition-like thoughts.

On the other approach, my swan concept is less like a word than like (perhaps very roughly) the dictionary entry for a word. It consists of representations that play a central role in guiding my use of the word; the swan concept, for example, might be a definition, or a prototype, or a theory that represents the knowledge that I use to classify organisms as swans and to make inferences about things I believe to be swans.

The first approach, then, identifies a concept with what I called in Section 2.4 the concept's "token structure," whereas the second approach identifies the concept with its "cognitive structure." Both ways of using the term "concept" are, I believe, just fine, as is a neutral way that does not commit to the nature of the concept itself but talks only of what its token and cognitive structure, respectively, accomplish. My ideal is the neutral way of talking, but at certain times—typically philosophical times—it is convenient to talk, as philosophers often do, of the concept while meaning its token structure, while at other times—typically psychological times—it is convenient to talk of the concept, as psychologists almost always do, while meaning its cognitive structure.

A few more words on cognitive structure. There are some theories of concepts that, if correct, offer obvious choices as to what should constitute a concept's cognitive structure. In the classical theory, to identify a concept's cognitive structure with its definition is irresistible; in the prototype theory, as you will see, it is natural to identify a concept's cognitive structure with its prototype. But in other theories, including my own, the boundary lines that delineate those representations central enough to serve as cognitive structure—separating them from those that, for example, constitute incidental beliefs about the corresponding category ("There is a swan in the meadow")—are less clear. That is not a problem: such theories do not presuppose a qualitative distinction between cognitive structure and the rest. Some representations are typically more and some less central to inferences involving the corresponding category ("typically," because epistemic context can make a difference as to which representations or beliefs decisively alter the course of reasoning). The line drawn by cognitive structure is nothing more than a loose and partial indication of this distinction of degree.

Psychologists working on conceptual change may have use for a coarser individuation of cognitive structure, capturing not a particular set of representations but a representation scheme. When the scheme corresponding to a certain set of words (the biological species terms, say, or the natural number terms, or the terms connected to thermal energy such as heat and temperature) undergo a major shift, conceptual change is said to have occurred. That implies a standard for individuating cognitive structure at the level of schemas; useful though it is for some kinds of discussion, I will not appeal to any such thing in this book. (I might add that this sort of individuation is typically accomplished by explicit sortals: a psychologist might distinguish, for example, "analog" and "numeral list" concepts of natural number [Carey 2009]. There is no need for token essentialism—no need to ask, say, of a certain infant's concept of the number two, whether it is essentially or merely incidentally an analog concept, and thus whether a change from the analog to the numeral list schema is a change of concept or merely a change in concept.)

Yet another notion of concept, which I should acknowledge before moving on, is the semantic notion introduced by Bolzano and Frege at the dawn of modern conceptual analysis (Section 2.2). I will ignore it, not because it is unimportant but because for the most part in this book the mind is where the action is, and the concepts that drive the dynamics of thought are in the first instance psychological things.

4.2 Representationalism

Thought proceeds by way of inference. It also proceeds by way of perception, memory, perhaps association, and analogy and metaphor (if these are noninferential). Inference, however, is paramount in the kinds of thought that are central to this book; other forms of mental activity are more or less taken for granted without being subjected to any serious examination. Almost everything I say will be in response to the following sorts of questions: What representations serve as the premises for such and such a conclusion? What is their epistemic status? What rule governs the inference? Is it reliable? Warranted? Finally, given the answers to these queries, what is the epistemic status of the conclusion itself? To that end, I conceive of thought as a rule-governed process by which representations are derived, inductively or deductively, from other representations.

It is expositorily helpful to talk about the inference rules in question as though they are themselves representations, directives written in some mental book of statutes that is consulted by the tiny inner logician who supervises our reasoning. But I don't believe that for a moment. Even the staunchest representationalist must allow that some patterns of reasoning are tacitly implemented, as psychological dispositions rather than as explicit rules, and as far as I can see this may be true for almost all of them, including the causal and other explanatory inferences that are especially important in this book.

Thus, though I may appear to assume, I do not genuinely suppose that the inferences I write about are mediated by explicit generalizations. Perhaps, as the "inferentialists" say, ordinary thinkers get from *Swanhood causes whiteness* and *Odette is a swan* to *Odette is white* without consulting any general belief about causality (Brandom 1998). What matters to me is not whether such an inference is "guided" by a general belief; it is that the inference fits a general pattern and that the pattern is in some sense an epistemically good one (reliable, or justified, or whatever). I may talk about the inference being made according to a causal logic, but this is simply a manner of speaking.

More generally, the model of mentation as occurring in a "language of thought," memorably championed by Fodor (1975), need not be absolutely true to be useful. Like many cognitive psychologists, I will make liberal use of the model, and of folk psychological notions such as belief, but you need not take everything I say literally. My principal interest is not in cognitive architecture but in epistemology—not in the implementation of inference but in its logical ramifications, in the epistemic status of its products, our conclusions about the nature of philosophical and other categories.

Two related idealizations are used freely in these pages. First, I frequently talk as though a reasoner's doxastic state is captured by a set of beliefs. That is far too crude to be true. Some things we are fairly sure of, some we are agnostic about, and some we think are clearly false. These gradations of belief are crucial to thought, especially in the course of theory construction, about which I will have plenty to say. A probabilist model in which every proposition is assigned a credence perhaps errs in the other direction, by demanding an exact level of confidence or subjective probability for every hypothesis that the thinker might entertain, but it does a far better job of capturing the subtleties of reasoning. It is ready as a supplement when needed.

Second, I assume that there is a one-to-one correspondence between natural-language words and the sorts of concepts I am writing about, that is, common natural kind concepts and philosophical concepts. Corresponding to our term "swan" there is a swan concept; corresponding to our term "water" there is a water concept; corresponding to our term "knowledge" there is a knowledge concept—where the concept has the same extension, inferential role, and more generally cognitive significance as the term. This is, I think, a harmless oversimplification. Among other benefits, it allows me to dispense with any awkward SPECIAL ORTHOGRAPHY for naming concepts; I simply make my claim about the natural-language term and let the one-to-one correspondence rule insinuate the parallel claim for the concept itself.

For similar reasons, I talk in the same breath about the "reference" of both natural language terms and natural kind concepts.

4.3 Case Judgments

Philosophical analysis begins with judgments about category membership—judgments as to whether something falls into the class of singular causes, or morally just acts, or instances of knowledge.

The term "judgment" suggests that our convictions about category membership are the end point of a reasoning process, by which I mean a process that is governed by some sort of logic (in a broad sense), that is in principle sensitive to our background beliefs, and that is somewhat accessible to conscious introspection. I will indeed conclude that the psychological processes that produce case judgments typically have all of these properties. But this is not a view shared by every philosopher.

Some writers have argued that the judging of cases is more like perception than like reasoning (Audi 2008; Chudnoff 2016). In this respect, case judgments are better labeled with the familiar term "intuition" (Bealer 1998).

In another view, case judgments are produced by reasoning, but that reasoning is "type 1" rather than "type 2" (or in the alternative terminology, "system 1" rather than "system 2"). Type 1 reasoning is said to be unconscious, automatic, effortless, and fast, among other things, whereas type 2 reasoning is conscious, controlled, effortful, and slow. If case judgments are the result of type 1 reasoning, then—as, for example, De Cruz (2015) argues—they are at least to some extent like perception.

These issues have been much discussed in the recent literature on philosophical methodology—see Cappelen (2012) and especially Machery (2017) for overviews—but I will not linger on the debates here. The reason is that many features of the psychology of case judgments are fixed by the nature of concepts, and so it makes sense to investigate concepts—the topic of the next two chapters—before committing to any particular view of categorization.

Nevertheless, I think it will be useful to say a little more about my picture of the workings of philosophical case judgments. (Machery [2017, §1.4] offers an extended defense of the view proposed here.)

Consider a scenario used by the psychologist Frank Keil to investigate children's natural kind concepts (the description is accompanied by a picture of a horse-like animal):

> These are animals that live on a farm. They go "neigh" and people put saddles on their backs and ride them, and these animals like to eat oats and hay and everybody calls them horses. But some scientists went up to the farm and decided to study them really carefully. They did blood tests and X-rays and looked way deep inside with microscopes and found out these animals weren't like most horses. These animals had the inside parts of cows. They had the blood of cows, the bones of cows; and when they looked to see where they came from, they found out their parents were cows. And, when they

had babies, their babies were cows. What do you think these animals really are: horses or cows? (Keil 1989, 305)

For myself, I think they are probably cows, though I am not certain (the typical adult response).

In making this judgment, I am not aware of invoking any general principles to make my decision, and so it seems "immediate" or "intuitive." With a bit of thought, however, I can recover the reasoning that went into the judgment. Whatever these animals are, something must have gone awry in the development either of their insides or of their outsides. What's outside, I know, is for the most part caused by what's inside. So I have a choice between unusual insides causing typical outsides or typical insides causing unusual outsides. The latter kind of scenario is more plausible; environmental and other factors often knock causal processes off course but far less often knock them back on course. Independently, there is just a lot more complexity to insides than to outsides; the deviation of insides from the species norm therefore seems more fantastical, far less likely, than the deviation of outsides. And so I might continue, with the reasons I give in retrospect supporting my initial judgment to a degree that is roughly in accordance with the confidence with which the judgment was originally made.[1]

Is this merely a post hoc rationalization that bears no connection to the actual categorization process? I do not take myself to have ruled out that possibility, but in this book I set it aside, taking it as a working hypothesis that categorization is inferential and that the logical structure of our categorizing inferences is largely, with some effort, at least partly psychologically accessible.

Why, then, when I make my judgment about Keil's scenario, am I initially unaware of the complexity of my reasoning? It cannot be that the inference is inferentially encapsulated or perceptual or otherwise completely unconscious: as I remarked above, I seem to have my reasons at my fingertips for classifying Keil's animals as cows, if I care to dwell on the case. (Keil's transcripts suggest the same of his young subjects.[2]) What happened in my brain when I made the classification is something that is much

---

[1] This observation is very different, I should note, from Deutsch's (2015) claim that philosophers habitually present explicit arguments to back their case judgments.

[2] Keil (1989, 166–174).

like conscious thought, except that I was not directly aware of it at the time. It is almost as though the inference was played out on the mental stage, but I was looking in a different direction.

This, I think, is indeed more or less what happened. The rationales for my categorizations are not submerged but merely overlooked—they are hiding in plain sight. If I go looking for them, I will find them, or much of them; I am therefore a rather reliable informant, if pressed, about the considerations that enter into my "intuition" of this thing's cowhood or that thing's knowledgehood. My reasoning toward the "cow" judgment in the Keil scenario, then, is somewhere between unconscious and fully conscious; it is, I am tempted to say, a "type 1½" inference: fairly fast, not particularly conscious, yet made not entirely without effort. The same is true, I suggest, of philosophers' judgments about knowledge in Gettier cases, causes in preemption cases, water in Twin Earth cases, and so on.

This picture does not rule out the occurrence of type 1 case judgments. Perhaps sometimes—say, when confronted without warning at dusk by a looming equine presence—I find myself thinking "Horse!" as a consequence of a wholly automatic and unconscious process. (Perhaps it is a perceptual process, if it is true that an animal's being a horse can be a part of the content of a visual experience [Siegel 2012].) Such judgments are, I think, of limited interest to methodologists of armchair philosophy. In any case, even they are in an important sense regulated by high-level thought and representation. I have in mind two forms of policing in particular. First, I conjecture, low-level categorization procedures are calibrated by beliefs about, or high-level representations of, the kind in question: the low-level rules that say "Horse!" in response to horse appearances are responsive to what we believe about horse appearances. The beliefs may not be consulted in real time, but they exert an irresistible influence in the long term.

Second, the verdicts produced by low-level categorization procedures can always be overridden by high-level cognition. Our final judgments in tricky cases thus reflect, directly or indirectly, the content of our beliefs or other high-level representations. Indeed, it is perhaps in order to make this regulation or supervision more effective that we are typically granted some degree of awareness of the grounds, if not the intermediate steps, of even the lowest-level case judgments. Seeing a horse bursting onto the road in front of me, for example, I have an inkling that my initial "horse" reaction is based on certain observed appearances and sounds. When I revisit my reaction, I can ask diagnostic questions about that basis: How reliable are appearances as a guide to species membership? Are insides or outsides more reliable?

These few remarks are enough, I hope, to give you some of the flavor of the way that I think about case judgments throughout this book: they are often fast and furious, even heedless, yet they are sufficiently of a piece with careful and sober reflection to merit the term "judgment" in preference to the grossly misleading "intuition."

4.4 Experimental Philosophy

Experimental philosophers use the methods of cognitive psychology and the other cognitive sciences to do philosophy. Often, their technique amounts to a democratic implementation of philosophical analysis, in which the case judgments of considerable numbers of ordinary people as well as professional philosophers are brought to bear, in the familiar way, on questions about the nature of intentional action, causation, and other philosophical categories (Knobe [2003] and Knobe and Fraser [2008] are two sterling examples among many). But sometimes they draw conclusions about philosophical methodology itself.

Much of this methodologically directed work concerns the psychology of case judgments. Its tenor is, as noted in Chapter 1, skeptical, casting doubt on the universality or reliability of case judgments and contesting the claims of analysis to yield philosophical knowledge. (Machery [2017] provides a comprehensive survey of experimental work up to the date of publication and marshals several lines of skeptical argument.)

One line of inquiry concerns the degree to which groups of people with different social or cultural backgrounds make the same judgments about the same cases. Experiments performed by Machery et al. (2004) and many follow-up studies suggest, for example, that under some circumstances East Asians tend to make different judgments from Westerners about Kripkean "Gödel" cases. For other thought experiments there are signs that gender, age, and personality have some influence on case judgments. (Machery's survey [2017] has the details.)

Do Westerners and East Asians, women and men, have different concepts of reference, knowledge, moral responsibility? Or are their judgments affected by philosophically irrelevant factors that tend to differ from group to group? The latter answer is suggested by the existence of considerable disagreement about cases within groups. Verdicts on Gödel cases, for example, have tended to show (very roughly) that about 60 percent of Westerners make "Kripkean" judgments, whereas only about 40 percent of East Asians do (Machery 2017, §2.1.1). That demonstrates a cultural or linguistic difference, but if anything an even greater individual difference within cultures.

Where do individual differences come from? One alarming possibility is that case judgments are systematically susceptible to transient influences that are irrelevant to the question at hand. Experimental philosophers have provided direct evidence that this is so. One sign of such influence is order effects, in which judgments change depending on the sequence in which thought experiments are evaluated. Some subjects' assessments of a "Truetemp" case, for example—a scenario in which you must decide whether a certain belief, formed in an unusual way, constitutes knowledge—were affected by whether they had previously been asked to make a judgment about a paradigmatic case of "non-knowledge" or a paradigmatic case of knowledge. Compared to the one "Truetemp" looked like knowledge; compared to the other not so much (Swain et al. 2008).

Another sign of inapposite transient influence is framing effects, in which apparently irrelevant changes in wording affect case judgments, and other indications of the power of context such as Valdesolo and DeSteno's (2006) finding that subjects were far more likely to judge, in a runaway trolley case, that it was "morally appropriate" to push the "fat man" on to the tracks to prevent the deaths of five others, if they had first seen a short comedy sketch.

Less transient but equally irrelevant influences on case judgments are personality and age: a large cross-cultural study of judgments about Gettier cases conducted by Machery et al. (2017a) found that laypeople are more likely to disqualify a Gettierized belief as knowledge (thereby conforming to the philosophical consensus) if they are more "open to experience," more "neurotic," less "conscientious," and older. (That does indeed sound like many epistemologists I know.)

To some extent, the experimental literature ought to serve to remind philosophical analysts of something that they already knew: many dialectically important case judgments are contested. Some armchair epistemologists judge the "Truetemp" belief to be knowledge even though most do not. Philosophers may often be found trying to talk one another into having the "right intuition," or when exhortation fails expressing frustration with "intuition-mongering" or the "clash of intuitions."

But surely there are some case judgments that are both exceptionally revealing and subject to almost universal assent. And surely these are the very same judgments concerning which the judgers are liable to experience "case certainty," such as the judgment that a Gettierized belief is not knowledge, or that the presence of a causal backup does not undermine the status of an actual singular cause. Some experimentally motivated skeptics accept the validity of these especially vaunted judgments while holding that they are too few in number to provide a substantial basis for philosophical analysis. Others cast doubt on even the most hallowed "intuitions." Turri (2016), for example, insinuates (without quite asserting) that judgments about even apparently straightforward Gettier cases are untrustworthy and that the accompanying feeling of case certainty is more a matter of peer pressure and professional solidarity than the accurate perception of a philosophically sure thing.[3]

My own modus operandi is, as I have said already, to inquire into the psychological origins of case judgments and the accompanying feelings of confidence before drawing conclusions about their reliability. Thus I will for the most part leave experimental philosophy unexplored until Chapter 10. There I will propose that different thinkers with identical concepts may make different case judgments because the logic of such judgments is inductive rather than deductive and there is a certain amount of plasticity, contextual sensitivity, and significant variation in expertise in human inductive reasoning (Sections 10.7 and 10.8).

Does this book belong at the experimental end of the philosophical bookshelf? Its emphasis differs from that of many armchair philosophy skeptics. Whereas the negative experimental philosophers have focused on

---

[3] Pushing back against this skepticism, Sosa (2009) proposes that diversity in case judgments may be due to different judges understanding thought experimental scenarios differently. In a more experimental vein, Wright (2010) presents data suggesting that the felt confidence with which case judgments are made increases with their stability—their resistance to order, framing, and other contextual effects. And Machery et al. (2017a, 2017b) have found considerable evidence for the stability and universality of Gettier case judgments, although nothing close to the virtually exceptionless assent that case certainty might lead an armchair jockey to expect.

cases where philosophical analysis may for systematic reasons break down, I am asking a more Cartesian question: how could it ever, even in the most favorable cognitive environment, succeed? The experimental philosophers and I are therefore working, as it were, opposite ends of the same question. We have a common interest, a common critical attitude to the practice of analysis (thinking that its reliability should not be taken for granted), and a shared belief that the empirical investigation of the mechanisms of thought is the most promising way to answer the question.

Nevertheless, this book is not experimental philosophy: it is a work of theoretical, rather than empirical, psychology, conducted in a rather speculative vein. By running ahead of the evidential tide, I hope to get a preliminary purchase on some deep questions about philosophical analysis with which experimental philosophy has yet to grapple: Why are we professional philosophers so confident about certain of our case judgments—a confidence that (speaking for myself at least) seems to be largely immune to the apparently subversive findings of the experimental philosophers? Why do we have the sense that, in testing philosophical hypotheses against these judgments, we are learning objective facts about the way things are—with respect to causality, morality, epistemology—outside our heads? In answering these questions, I will not succeed in preserving entirely the idealized picture of philosophical analysis as allowing a priori access to a Platonic world that exists independently of our thought. But I do hope to show that, when circumstances are right, we can from the sitting position attain genuine philosophical knowledge.

CHAPTER 5

Natural Kind Concepts

On to the psychology of concepts of philosophical categories, then? To the psychology of the concept of knowledge, of the concept of cause, of the concept of agency? Not just yet. There is no well-developed, general, psychological theory of the philosophical concepts, so I will look elsewhere for inspiration, to the concepts that psychologists know best—namely, the concepts of what I will call *basic natural kinds*. These are the lower-level categories of chemical substances, such as gold and water, and biological taxa at approximately the genus level—what are called *folk genera*—such as horses, apples, and raccoons.[1] This list can perhaps be fruitfully expanded, but I will limit myself to chemistry and biology. In using the term "natural kinds," note, I am merely conforming to a long-standing naming convention, rather than attributing to these categories any special metaphysical significance. The same goes for "basic": it is borrowed from cognitive psychologists' notion of a basic level of categorization, and is invested with no philosophical implications.

A short history of the psychology of the folk genera will set the scene for a close examination of what is now, if not the dominant view of folk genus

---

[1] The folk genus level of psychological classification corresponds roughly to those groups that ordinary people refer to as "species"; it does not always correspond to the genus level in the Linnaean hierarchy. The notion of a folk genus, then, belongs to psychology, not to biology. The correspondence between folk genera and scientific genera is nevertheless much closer than you might expect, at least for the vertebrates and for many plants (Berlin et al. 1974; Atran 1990).

and other basic natural kind concepts, certainly the view with the most momentum: psychological essentialism. If the psychological essentialist theory of concepts is correct, then definitions, reference-fixers such as referential intentions, and essence postulates play almost no cognitive role in reasoning about basic natural kinds, and in particular, they play almost no role in classification procedures, that is, in the chains of inference leading to conclusions about which specimens belong to which kinds. This leads me in the next chapter to a view of basic natural kind concepts that I call causal minimalism, which posits a structure for the concepts that is the essentialist structure pruned of vestigial metaphysical commitments.

5.1 Locke and Classical Empiricism

A natural kind concept, John Locke writes in his *Essay Concerning Human Understanding*, is "nothing else but a collection of a certain number of simple ideas, considered as united in one thing." For example,

> The idea which an Englishman signifies by the name *swan* is white colour, long neck, red beak, black legs, and whole feet, and all these of a certain size, with a power of swimming in the water, and making a certain kind of noise, and perhaps, to a man who has long observed those kind of birds, some other properties, which all terminate in sensible simple ideas, all united in one common subject. (II.23.14, p. 305)

What Locke articulates is of course an instance of the early modern molecular approach to complex concepts, according to which they are psychologically composed of less complex concepts, and ultimately of basic concepts, that is, Locke's simple ideas. Because the simple ideas literally constitute the swan concept, to apply the swan concept to a specimen just is to apply, jointly, the simple ideas to the specimen. Likewise, to think any other thought using the swan concept is to think the same thought using the conjunction of the simple ideas. Consequently (so the story goes), the conjunction functions like a definition: satisfaction of the simple ideas is necessary and sufficient for swanhood.

5.1. LOCKE AND CLASSICAL EMPIRICISM

Let me peel apart two distinct aspects of the Lockean theory of natural kind concepts. First, such concepts are, according to the theory, functionally equivalent to lists of necessary and sufficient conditions, hence to definitions. This thesis, applied to concepts generally, is the classical theory of concepts discussed in Section 2.4. Second, the conditions in question all involve, ultimately, sensible qualities. This is concept empiricism. Locke's account of natural kind concepts, then—and apparently of all other concepts—is classical empiricism.[2]

Every Antipodean philosopher knows how Locke's theory falls short. For a Lockean Englishman, a swan is white by definition. But when Westerners observed specimens of *Cygnus atratus* for the first time in Australia they identified them as black swans.[3] Such a classification would be a psychological and logical impossibility, if Locke were right.

One way to defend the Lockean approach against the black swan is to hypothesize that upon first coming across *Cygnus atratus*, we redefine our swan concept on the fly to allow for black swans. Another perhaps more feasible defense is to assert that whiteness cannot ever have been, in spite of Locke's suggestion to the contrary, a part of the mental definition of swan. Both moves are uncomfortably ad hoc.

Psychologists have in any case largely given up on the search for a mental definition of "swan" and other natural kind terms (see Section 2.4). One reason is the difficulty of finding definitions that survive experimental scrutiny; another, historically more important reason is that an alternative theory of concepts promises to explain many more elements of the psychology of categorization. That is the prototype theory of concepts.

---

[2] This exegesis is complicated by Locke's view that ideas may be built through operations of relation and abstraction. The thesis that ideas so formed can be captured by definitions couched in primitive sensory terms fits well with the Lockean project, yet relation and abstraction are not sufficiently fleshed out in the *Essay*, I think, to guarantee its truth. That natural kind concepts are conjunctions of simpler and more or less sensory ideas is something Locke repeats many times, however, and that is more than sufficient for my purposes here.

[3] The first Europeans to observe the black swan were Dutch, part of a 1696 expedition to Western Australia under the command of Willem de Vlamingh. The ship's journal and the report to the directors of the Dutch East India Company identify the birds explicitly as swans: "On the 7th [January 7th, 1697] the whole of the crew returned on board with the boats, bringing with them two young black swans" (Major 1859, 123).

5.2 Rosch and the Prototype Approach

Typicality effects: this is the blanket term for an array of phenomena in categorization and related tasks demonstrated by Eleanor Rosch, her collaborators, and other investigators in the 1970s.

When subjects—typically college students—attempt to decide as quickly as possible whether a given organism (usually presented in pictorial form) is a bird, they are quicker on typical birds, such as robins, than on atypical birds, such as ostriches or penguins. Indeed, birds can be assigned a quantification of their typicality that predicts speed of categorization. The same typicality scale predicts many other psychological phenomena: subjects' own ratings of the specimens' typicality; subjects' willingness to extrapolate a specimen's properties to other birds (the more typical the specimen, the more enthusiastic the extrapolation); asymmetry in similarity ratings (an atypical specimen is rated as more similar to a typical specimen than vice versa); and more (Rosch and Mervis 1975; Rosch 1978).

To account for typicality effects, Rosch proposed what has become known as the prototype theory of concepts—or better, the prototype approach, since there are many variants and specific computational models based on Rosch's ideas. Just one way of implementing the prototype approach will be considered here, an instantiation that I will call "the" prototype theory, but the lessons generalize.[4]

In the prototype theory, as in the Lockean theory, a concept might be thought of as specifying a list of observable properties. The prototype theory's rule for using the list to determine category membership is rather different, however, than the Lockean rule. None of the properties is represented as strictly necessary for category membership; what matters is that a specimen should have sufficiently many of the properties on the list, in a sense to be spelled out below. If it does, it is classified as a member of the category; if not, as a nonmember. When used to determine category membership in this way, the property list ceases to be a definition and becomes a *prototype*, a schema that captures the relevant properties of a

---

[4] For the variants and sophistications of the prototype theory, and in particular for the version—or close rival—called exemplar theory, see Smith and Medin (1981) and Murphy (2002).

quintessentially typical or representative member of the category. Category membership is determined, in effect, by assessing a specimen's similarity to this prototype.

There is both a quantitative and a comparative aspect to similarity. Different properties in the prototype may be differently weighted, so that possession of some counts for more toward category membership than possession of others. The procedure for determining category membership computes an overall degree of similarity between a specimen and the candidate prototypes, using the weighted properties. A specimen is then classified as a member of a category if, first, its similarity to the category prototype exceeds a certain threshold, and second, it is not more similar still to a prototype for a competing category. (Two categories compete if a specimen cannot be a member of both. "Swan" and "goose" compete. "Swan" and "bird" do not compete; nor do "goose" and "Christmas dinner." When classifying a specimen as a member of a folk genus, then, the competing categories are in the first instance the other folk genus categories.)

Let me pause for a moment to consider the ways in which the prototype theory might explain the travails of modern conceptual analysis (see also Section 2.4). As Stich (1993) and Ramsey (1998) point out, philosophical categories whose essential natures are determined by prototype concepts will tend not to yield to conceptual analysts who expect a simple list of necessary and sufficient conditions to capture the ultimate criterion for category membership. For related reasons, categories with "prototypical" essential natures will not spawn simple analytic truths, but rather clusters of complex disjunctive consequences. (If the swan concept is a prototype, then swans must have some cluster or other of swanlike properties; each possible cluster constitutes a term in a long disjunction spelling out this conceptual necessity.)

The prototype theory is also able to provide a framework for explaining various phenomena uncovered by experimental philosophers. Order effects might, for example, be caused by the transient influence of salience on the weighting scheme used to determine category membership.

In spite of its explanatory virtues, the prototype approach is fading as a model for concepts of natural kinds, and in particular of basic natural kinds such as swan or beech. Consider the swan concept. Locke's list of features—white color, red beak, trumpeting, with a power of swimming,

FIGURE 5.1. Raccoon (left), transformed (right). From Keil (1989, 177), © 1989 Massachusetts Institute of Technology, reprinted with permission of MIT Press.

and so on—might make just as good a swan prototype as a swan definition. Suppose, then, that "the idea which an Englishman signifies by the name *swan*" is characterized by this Locke-like prototype, including the white color.

Such a swan concept will be applied to Australasia's black swans even though its prototypical swan is white if, as seems likely, a black swan is in spite of its color more similar to the swan prototype than to the prototype for any other folk genus. In that case, Europeans with such a concept would recognize *Cygnus atratus* as a variety of swan, as indeed they did.

Other more extreme cases, however, present greater difficulties for the prototype theorist. They fall into two classes: cases of extremely atypical individual specimens, and cases where the initial conception of a basic natural kind turns out to be grossly inaccurate.

Keil's (1989) "transformation" experiments provide compelling examples of the first sort. Children of various age groups—roughly 5, 7, and 9 years old—were tested by Keil and his collaborators using the following paradigm. Each child was shown (to take one example) a picture of a raccoon, identified as such. They were told that the raccoon was later subjected to a certain cosmetic transformation: its fur was trimmed, it was dyed black except for a broad white stripe down its back, and it had a sac of "super smelly odor" implanted in its body. A picture of the end product—visually indistinguishable from a skunk—was then displayed (Figure 5.1), and the children were asked whether the animal was still a raccoon or was now a skunk.

Younger children tended to reply as the Lockean or prototype theory predicts: because the animal lacked the characteristic observable properties of a raccoon, but had those of a skunk, it was now a skunk. Older children

5.2. ROSCH AND THE PROTOTYPE APPROACH

and adults tended to reply that it was still a raccoon. This latter response is impossible to square with the prototype theory as it is usually understood: the transformed raccoon is far closer to fitting the skunk than the raccoon prototype.

Another sort of case, closer in epistemic structure to that of the black swan, is well known to philosophers from the work of Kripke, Putnam, and others. It seems possible that the prototype connected to a basic natural kind concept might radically misrepresent the typical appearance and behavior of members of the kind. Consider, for example, a class of cases that is less outré than Putnam's (1962) "robot cats" or Kripke's (1980) "blue gold" and that has had some influence among psychologists of concepts. Certain species are radically sexually dimorphic: the males and the females do not resemble one another very much in either appearance or behavior. The male in some species of barnacles, for example, such as *Trypetesa lampas*, is a microscopic parasite living on the female. Suppose that when *T. lampas* is first observed, only the females are known. The corresponding prototype will then reflect female appearances and behavior; the males will fit the prototype quite badly (and may well better fit the prototype for some other organism). But on discovering the minute males cradled in the females' conjugal embrace, we have no hesitation—once we understand their reproductive role—in counting them as members of *T. lampas*, thus as falling under the *T. lampas* concept. Their failure to fit the prototype seems irrelevant to this act of categorization (Murphy 1993, 190). The same is true for animals that undergo significant metamorphosis (Rips 1989).[5]

One response to the dimorphism case mirrors a potential Lockean response to the black swan mentioned above: the biological discovery, a prototype theorist might hypothesize, causes an on-the-fly conceptual reconfiguration, a revision of the prototype itself. This suggestion is ad hoc and to a great extent begs the question: Why would you want to reconfigure the prototype to include the newly revealed appearances and behavior unless you had already decided that the novel creatures in question belonged to the corresponding kind?

An alternative Lockean response is more promising: perhaps the difficulties with Keil's transformations and Kripke's discoveries are caused not

---

[5] For some complications, see Hampton et al. (2007).

so much by the hypothesized prototype structure of the concepts as by the empiricist assumption that the properties appearing in the prototype are predominantly observable properties such as color, shape, and overt behavior.

Empiricism has usually gone hand in hand with the prototype theory, in part for historical reasons and in part because typicality effects are strongly predicted by empiricist prototypes, that is, by a typicality structure built around appearances and behavior. But the connection might be weakened. Work by Ahn (1998) and Rips (2001), for example, can be read as an attempt to refresh the prototype theory by adding "deep" or theoretical properties, such as internal structure or DNA signatures, to the prototypical mix—properties that are presumed to be shared by black and white swans, regular and gussied-up raccoons, and male and female members of *Trypetesa lampas*.

Another approach thinks of empiricist prototypes as constituting only one part of a basic natural kind concept, with the other part or parts being what determines that specimens badly fitting the prototype nevertheless belong to the category. An example of this approach is Osherson and Smith's (1981) suggestion that concepts house both a prototype and a "core" that acts rather like a definition. The prototype, which may be mostly or wholly composed of observable properties, acts as a categorization heuristic; it is the operation of this heuristic that is responsible for typicality effects. The core, which cites deeper properties as the ultimate determinants of category membership, is what dictates the category membership of cosmetically transformed raccoons or parasitic male barnacles.

What is in the core, then? Or what deep properties need to be incorporated into prototypes to explain Keil and Kripke classifications? That is not an easy question to answer. Suppose that some deep property or property complex P is a part of the core or the prototype of my swan concept. It seems to be possible, for any P you like, to imagine discovering that, as a matter of empirical fact, some or all swans lack P: we were simply mistaken in thinking that P was an important component of swanhood. What ought to be logically proscribed—the classification of a P-less specimen as a member of the category—in fact turns out to be entirely possible. The theory of natural kind concepts that comes closest to explaining this phenomenon is psychological essentialism.

5.3 Psychological Essentialism

In the Lockean view, when an organism is classified as a member of a basic natural kind on the basis of its observable properties, it is because the specimen satisfies a definition. In the prototype view, it is rather because the specimen resembles a prototype. According to a third approach to concepts and categorization, the *theory-theory*, classification involves scientific or proto-scientific reasoning: we represent a theory about the relation between category membership and observable properties, and we use the theory to determine the likelihood that a specimen with such and such properties is a member of such and such a category (Carey 1985; Murphy and Medin 1985; Medin and Wattenmaker 1987; Gopnik 1988). Psychological essentialism is an application of the theory-theory approach to folk genera and other basic natural kinds, according to which we are naturally disposed, when constructing causal theories about categories of this sort, to give them an essentialist structure.

The psychological essentialist theory of basic natural kind concepts, perhaps first formulated under that name by Medin and Ortony (1989), is built around the following two theses (Gelman 2003):

1. Humans represent every basic natural kind as having an essence, a property necessary and sufficient for (and indeed, constitutive of) membership in the category. They need not, and often do not, represent the nature of the essence.
2. The essence of a kind is represented as causing its members' characteristic observable properties. For example, the swan essence is represented as causing swans to grow white feathers and a red beak and to engage in their characteristic swan behaviors, such as trumpeting and swimming.

Humans may represent the causal powers of the essence, then, without representing any of the essence's intrinsic properties. They know that something does the causing, and that the something is the ground of category membership, but they do not (or need not) have a view as to the nature of that something.

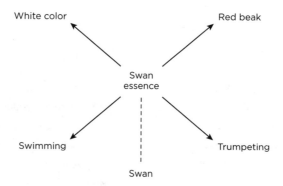

FIGURE 5.2. An essentialist theory of swans.

Let me put this in pictorial form. For each of a thinker's basic natural kind concepts, psychological essentialism attributes to that thinker a causal theory having the "starburst" structure shown in Figure 5.2. The arrows in the figure represent causal relations; for example, the arrow between swan essence and white color represents the thinker's belief that the essence causes swans to develop a white coloring. The dashed line represents a metaphysical relation: to be a swan is to have the swan essence, or (perhaps equivalently, and as the philosophical notion of essence implies) the swan essence is identical to the property of swanhood. A number of variants of the essentialist theory may be distinguished. Some of these are described in Strevens (2000); in this discussion, however, I will focus on what I consider to be the pure form of the type.

Essentialism is so called because it attributes a belief in essences to the naive human reasoner, not because its advocates among psychologists themselves endorse an essentialist metaphysics for biological taxa and so on. The theory shown in Figure 5.2 is therefore a "naive" or "folk" theory, meaning a theory that we humans are naturally inclined to build, but which may later be replaced by a more advanced or more scientific successor.

A naive theory need not be literally naive; rather than replacing it we may augment it with new insights and new information without altering its fundamental explanatory structure. We then end up with a rather sophisticated naive theory. That is what has happened, most likely, with belief/desire psychology: in our dealings with one another, we are all still "folk psychologists," but because this folk psychology enables us to reason in extremely

5.3. PSYCHOLOGICAL ESSENTIALISM

subtle and successful ways about the course of human thought and social relations, it has been recruited as the basis for numerous scientific models of thought (including the models of concepts discussed here). Similarly, the "folk biology" of people living in close proximity to the natural world has, in the case of plants and higher animals, a taxonomical structure that mirrors that of the Linnaean system (see note 1).

To infer that a certain specimen is a member of a basic natural kind, the essentialist thinker typically calls upon the causal beliefs linking essence and observable properties—which I will call the *core causal beliefs*—in conjunction with the standard apparatus of causal inference. For example, because they believe that certain appearances and behaviors are the causal consequences of the essence of swanhood, they will be disposed to infer that a bird exhibiting these appearances and behaviors is a swan, in the absence of any other likely causes. This is an application of a familiar tenet of causal logic: if possession of F causes G, and x is G, then provided that there are no other plausible causes of x's G-ness, infer that x is F.[6] Since causes are explanations, such an inference might be thought of—optionally—as a kind of inference to the best explanation.[7]

The essentialist's understanding of categorization as inference to the most likely cause or the best explanation should be distinguished from the sort of view—in effect, a version of the prototype view—in which we categorize something as a swan because that thing's causal structure is a sufficiently good match to the paradigmatic causal structure articulated by our theory of swans. In the essentialist view (and as you will see, in my own

---

[6] Provided, of course, that nothing tells independently against x's F-ness.

[7] There is a certain imprecision in the claim that the essence is a cause of each of the characteristic properties. Does it imply that external conditions needed for the production of white feathers are included in the essence? Does it imply that the causation of white feathers and red beaks, or at least the parts of the process that occur within the swan, are from beginning to end identical? No to both questions, surely. An interpretation that is neither too weak nor too strong might be as follows: the internal parts of each of the characteristic property–causing processes at some point entirely overlap and the point of overlap is or includes the essence. This allows for a single, internal essence to be triggered in different ways (and presumably at different times) for different properties, while requiring that the triggering of that essence is the sole initiator of the presumably somewhat different causal chains leading to the expression of each of the characteristic properties. Perhaps also, the essence should be understood to play a guiding as well as an initiating role. It is vital to the development of essentialism, I think, to flesh out these causal commitments, as they play a crucial role in steering the causal inferences that give psychological essentialism much of its explanatory power.

view also) the causal beliefs guide the inference that constitutes the categorization, while in the structure-matching view, they function merely as points of comparison, just as you might compare colors or sounds. Consequently, in the essentialist view the beliefs alone (inductively deployed) drive categorization, while in the structure-matching view, an additional principle is needed: "A bird x is a swan just in case its causal structure is sufficiently similar to the paradigmatic swan causal structure." (I take the inferential approach to categorization as central to the theory-theory of concepts, although many psychologists sympathetic to the theory-theory seem to assume, if only implicitly, a structure-matching view.)

Note also that in the essentialist view, the inductive application of the core causal beliefs is far from being the only way to draw the conclusion that some bird is a swan. You might learn of its swanhood from an ornithological authority, or you might infer it from your knowledge that the only birds in the vicinity are swans (if you catch just a glimpse of the specimen). Essentialism, and the theory-theory more generally, does not impute any special authority to causal inference by comparison with these other kinds of inference to swanhood. It claims at most that inference using core causal beliefs is the way in which we usually infer a specimen's kind membership from its possession of the kind's characteristic observable properties.

The essentialist theory of basic natural kind concepts is well able to explain a range of striking patterns of inference and categorization in both children and adults; this evidence in its favor is surveyed by Strevens (2000) and Gelman (2003).[8] Of special interest is its ability to account for our disposition to classify Keil's transformed animal as a raccoon.

In the Lockean and prototype theories, appearance and behavior are everything. If an animal looks and acts like a skunk, it is a skunk; if, by contrast, it is a raccoon, then it looks and acts like a raccoon. The essentialist theory, too, imputes great diagnostic power to basic natural kinds' appearance and behavior, but defeasibly: the link between, on the one hand, the essential nature of raccoonhood or skunkhood, and on the other hand, characteristic raccoon or skunk appearances and behavior, is causal, and

---

[8] A promising exercise, which I have not seen carried out in print, is to canvass the ways in which essentialism might explain the typicality effects.

thus can be broken or superseded. A raccoon might therefore be made to look like a skunk while retaining its essential nature as a raccoon.

How to tell the difference? Simply looking won't help, but if you know something about the causal origins of an animal's observable features, you can make sophisticated causal inferences. Suppose that you happen to know that an animal's appearances and behaviors were imposed by an external actor. Then you can infer that they were not caused by its essential nature. But if they were not caused by its essential nature, then they are not reliable indicators of that nature, and so not reliable indicators of its species. If you know, for example, that an animal's skunky aspect was created by a cosmetician, then you should not regard that aspect as evidence for skunkhood, or for any change in the animal's essence at all. If it started out as a raccoon—if it started out with the raccoon essence—it surely still has the essence, and so it is surely still a raccoon. (Clinching the inference is the knowledge, or at least the supposition, that the animal has not been subject to any transformations other than the cosmetic job).[9]

What about Kripkean revelations, such as the discovery that the male of the barnacle species *Trypetesa lampas* looks quite different and behaves very differently from the female? The key to accommodating such epistemic trajectories is to allow that existing core causal beliefs can be refuted, and new core causal beliefs introduced, by empirical inquiry. Core causal beliefs are simply the thinker's ordinary beliefs about the causal consequences of possessing the essence, open to revision like any causal belief in the light of new evidence.

To see how core causal beliefs might change, consider a toy example (toy so that the epistemic context can be specified precisely). One fine day you come across a curious group of waterbirds that fit none of your existing folk genus concepts. They have pink feathers, blue beaks, a distinctive trombone-like call, and oddly they have no interest in or facility for swimming. You formulate a new concept corresponding to what you (correctly)

---

[9] Why do the younger children in Keil's experiments fail to classify the animal as a raccoon? Keil proposes that they have resemblance-based rather than causation-based concepts of folk genera and other basic natural kinds. I prefer an explanation in which young children do in fact have causation-based concepts, but they lack the inferential firepower to follow the consequences of their causal theories to their logical conclusions (Strevens 2000, §3.3).

suppose is a new kind of bird, and a new term to go along with that concept: you call the birds *schwanns*.

If psychological essentialism is correct, then when you form your schwann concept, you construct a theory of schwanns which, first, posits that there is a property necessary and sufficient for schwannhood, the schwann essence; second, that the schwann essence causes the growth of pink feathers; third, that it causes the development of a blue beak; and so on, adopting a core causal belief for each of the properties you take to be distinctive of schwanns.

You regard these beliefs as mere hypotheses. If it turns out that someone sprayed the schwanns you saw with pink paint, and that they are in fact green underneath, you will replace the belief that the schwann essence causes the growth of pink feathers with the belief that it causes the growth of green feathers. If it turns out that the birds you saw were all males, and that the females have green feathers, then you will replace the belief that the schwann essence causes the growth of pink feathers with the belief that it causes the growth of feathers that are either green or pink depending on sex. A series of discoveries of this sort might in principle result in your abandoning every one of your original core causal beliefs. The kind of Kripkean discoveries that cause problems for the Lockean and the prototype theorist are deftly handled, then, by essentialism.

⌇

There is a clear affinity between the psychological essentialist theory of basic natural kind concepts and some aspects of Kripke's and Putnam's ideas about reference-fixing (Putnam 1975; Kripke 1980). Putting things rather loosely, Kripke and Putnam suggest that when a basic natural kind term such as "gold" or "water" or "cat" is first introduced, its reference is fixed by a stipulation that the term refer to all specimens of the same kind as some paradigm sample (some trove of gold or local population of cats, for example). The intender in question may know little or nothing of the nature of the kind in question, but "same kind as" is supposed to have some determinate meaning in their minds; between the baptismal sample and this relation of sameness of kind, an extension for the term is fixed without

5.3. PSYCHOLOGICAL ESSENTIALISM

the fixer knowing what common invisible property—what essence—binds together the members of the extension.

What unites the members of a psychological category, according to psychological essentialism, is the same sort of unknown property. Rather than transparently spelling out the category's ground or essential nature, the thinker simply commits to there being some such ground. Although they may represent nothing about the intrinsic nature of the ground, they give their supposition empirical significance by positing causal connections between the ground and certain observed properties of the category. In the essentialist story there is no explicit mental reference to a baptismal sample, but the observable properties that trigger concept formation are, as in the schwann example above, typically suggested by the characteristics of some limited population.

Psychological essentialism and the Kripke/Putnam approach to natural kind terms, then, make similar assumptions about our basic natural kind cognition. They are theories of different things—the former is primarily a theory of categorization, while the latter is primarily a theory of reference—so the correspondence is no more than thematic, but the two sets of ideas point in a broadly similar direction.

Let me remark on two other appealing features of psychological essentialism before I proceed to its refutation.

The first is that essentialism shows, far more clearly than the prototype theory, how a theory of concepts can be psychologically compositional without being semantically compositional. A psychologically compositional theory is one according to which concepts, or rather concepts' cognitive structures, are built largely from other, psychologically prior concepts. Psychological essentialism fits that bill: an essentialist concept's cognitive structure takes the form of a theory built from concepts of relatively observable properties, namely appearance and behavior, along with the concept of causation. (The theory also involves a token of the kind concept itself. You might worry that this confronts any anticipated compositional theory of concept acquisition with a bootstrapping problem, and you would be right; I take up the matter in Section 15.1.)

In the early modern molecular theory of concepts, psychological composition goes along with semantic composition: a concept's psychological components are also its semantic components, thereby participating in its definition. Something of this picture lives on today in many philosophers' thinking, I believe—perhaps most notably in Fodor's (1998) argument that (simplifying for expository purposes) because the great majority of concepts do not have definitions, they must be "atomic," where I understand Fodor's atomicity to consist in a lack of psychological structure.

But the psychological components of an essentialist concept—the concepts that make up the corresponding essentialist theory—are not (or at least, need not be) semantic components in any sense. They do not function as parts of a definition or as a reference-fixing stipulation but rather as ordinary empirical posits or beliefs, subject to the usual dynamics of inductive reasoning. I will have much to say about this in the coming chapters.

My second remark is that essentialist concepts are especially well positioned to pick out categories that have the elusive character of "substantiality," of having some objective standing or significance in the natural world. The reason is that their cognitive structure—an empirical theory—behaves upon contact with the world like the "molten wax" of an open concept, taking on the imprint of the world as core causal beliefs are acquired, discarded, or revised in response to the empirical evidence (see Section 1.3). Thus, an essentialist concept adjusts itself to reflect the contours of the relevant physical (or chemical, or biological) subject matter, and so tends to pick out a category that reflects those contours, that is "natural" rather than "artificial" or "imposed"—a natural kind.

5.4 Against Essentialism

The psychological essentialist account of basic natural kind concepts is a theory that is able both to accommodate philosophical subtleties and to explain a wide range of psychological phenomena. It has, nevertheless, certain weaknesses, and these weaknesses are connected to the essences—or rather, the representations of essences—that give it its name.

Let me begin with a discovery scenario that, unlike the Kripkean discoveries discussed above, psychological essentialism seems unable to explain. Essentialism, as I have said, treats the core causal beliefs as ordinary empirical hypotheses, subject to confirmation or disconfirmation. But a commitment to the existence of an essence that plays a certain causal role is built into basic natural kind concepts, if essentialism is correct.

The commitment can be stated as follows, for a given category K:

There exists a property P (the essence of K) and properties $C_1, C_2, \ldots$ (the characteristic observable properties of K) such that

1. Possession of P is necessary and sufficient for membership in K, and
2. Possession of P causes C_1, C_2, and so on.

Because of the existential quantification over P and the C_is, this commitment does not specify any intrinsic properties of the essence and does not identify the characteristic properties. The thinker may well specify some of the former and will almost always putatively identify some of the latter, but these specifications and identifications are not a part of the fixed commitment that comes with every essentialist concept. The commitment is, rather, more abstract; it is to the existence of something that plays what I will call the essence role, captured pictorially in the "starburst" structure of the naive theory shown in Figure 5.2.

What is the force of the commitment? In what sense, according to psychological essentialism, is the belief fixed that there is for any kind something that plays the essence role? Essentialism does not entail that the commitment is stipulated by the thinker, or that it is a conceptual truth for any other reason. What it does entail—insofar as it claims to be a general theory of basic natural kind concepts in humans (or modern humans, or post–Agricultural Revolution humans, or some other class of cogitating life-forms of which we are all members)—is that the essentialist commitment is psychologically fixed: this is how humans conceive of basic natural kinds. Of psychological if not semantic necessity, then, when humans think about basic natural kinds, they assume that something with the essence role is pushing the buttons and pulling the strings.

If this were true, it would be impossible, in the psychological sense of possibility, to discover that there is no property that plays the essence role in swans: to do so, you would have to assent to the hypothesis *Swans have no essences*, but your only means for doing so is a concept of swan that is a theory that explicitly and permanently ascribes essences to swans. You would be, in effect, asserting a contradiction. Of course, we could discover that there is no such property in the animals we took to be swans, but it would follow that they are not in fact swans, since the presumption that there exists such a property is lodged immovably in the swan concept itself.

Arguably, however, we have made just such a discovery. Many biologists and philosophers of biology would say that there is no single property that is shared by all swans and that is causally responsible for most or all characteristic swan appearances and behavior. There is, rather, a distinct mechanism responsible for each characteristic property, and parts of any such mechanism may be missing in any given swan. Conversely, if there is a single property or property complex that is necessary and sufficient for swanhood (a question concerning which there is much debate), it does not occupy the central role in the causation of the characteristic appearances and behaviors that a psychological essentialist concept assigns to the essence (Devitt 2008). Yet still there are swans.

These doubts about essences are largely correct, I think, as a matter of biological fact, but even if they were not, it would hardly matter; what is important is that they articulate a coherent possibility. Psychological essentialism would seem to imply otherwise: it would seem to imply that, given the inflexibility of our swan concept in the matter of the essential role, either there is a property that plays such a role in swans, or by our lights there are no swans (since we cannot coherently conceive of there being swans). Essentialism fails, then, to predict our actual reasoning about species-level taxa: we abandon some or all of the essentialist suppositions about the taxa but retain the taxa themselves. To put it another way, we react to the falsehood of essentialism not by abandoning our naive theories and therefore our concepts of the folk genera, as we would have to if they were inherently essentialist, but by augmenting them with new anti-essentialist beliefs.

How might the psychological essentialist defend their theory against this objection? Perhaps psychological essentialism need not mandate that the swan concept—that is, the naive theory of swans—is, in its fundamental explanatory structure, unchangeable. Perhaps essentialism is committed only to there being a "natural tendency" for us to think that in any kind, there is something that plays the essence role. Such a commitment has psychological momentum, but it is not unstoppable. So when faced with overwhelming biological evidence, it is possible for the swan concept to adjust itself to fit.

This response is compatible with the nondoctrinaire mien of many psychological essentialists and their fellow travelers (Medin and Ortony 1989; Gelman 2003; Cimpian and Salomon 2014). But it constitutes an implicit renunciation of essentialism's ambition to be a general theory of basic natural kind concepts. The general theory, it seems, will accommodate, but will not insist on, the belief that there exist properties playing the essence role. Further, it will explain how such a belief can rationally enter into or smoothly depart from our thinking about the kinds, and it will explain why even biologically sophisticated individuals make the same sorts of inferences about disguised raccoons as those committed to the existence of essences—not to mention the other kinds of inferences that have been touted as evidence for essentialism but which are just as prevalent in those thinkers who have embraced the antiessentialist consequences of modern biology. This more general theory is the correct theory of basic natural kind concepts; psychological essentialism is by its own concession something strictly more parochial.

A quite different argument against the psychological essentialist theory of basic natural kind concepts was first mounted by Malt (1994) and then reprised in Strevens (2000, §5.3). The objection looks closely at the inferential dispositions of thinkers who have (what they believe is) some information about a basic natural kind's essence. Whereas psychological essentialism predicts that this information will be central to the thinker's categorizations, it in fact appears to play little or no role.

The basic natural kind in question is water. Malt's aim was to test the hypothesis that ordinary people with some knowledge of chemistry regard H_2O as the essence of water. She asked her subjects to estimate the percentage of H_2O in common liquids such as seawater, tea, blood, sweat, tears, swamp water, and mineral water. The estimated percentages showed surprisingly little connection to the subjects' judgments as to which of the liquids were kinds of water and which were not.[10] Tea, for example, is rated

---

[10] Malt did not experimentally verify her judgments as to what her subjects would and would not count as a kind of water, but my own informal research leads me to believe that her assumptions are largely correct.

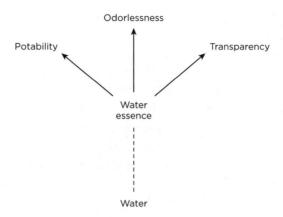

FIGURE 5.3. An essentialist theory of water, before learning about H_2O.

as containing more H_2O than seawater and far more than swamp water, but the latter two are categorized as kinds of water while the former is not.

This poses a serious problem for the essentialist theory of concepts. Consider a thinker who has an essentialist concept of water but knows nothing about H_2O. Their concept has the theoretical form shown in Figure 5.3, which represents the existence and the causal role of the water essence but nothing about the essence's intrinsic nature.

Now suppose the thinker goes to school. There they learn that water has its characteristic observable appearances and behaviors because it is composed of H_2O. It is H_2O, in other words, that causes water's potability, its odorlessness, its transparency. According to the essentialist theory that is their concept of water, the thing with this causal role is water's essence. They ought to reason, then, that the essential nature of water—the thing that makes a substance a kind of water—is H_2O. They will therefore enrich their water theory by substituting something like "being largely made up of H_2O" for the placeholder representation "water essence," as shown in Figure 5.4.

To have a theory with this structure is to believe that "being largely made up of H_2O" is necessary and sufficient for a specimen's being water. If you count swamp water as a kind of water, then, you should count anything made up of the same or a greater proportion of H_2O as water. But Malt's data show that this prediction of psychological essentialism is quite wrong.

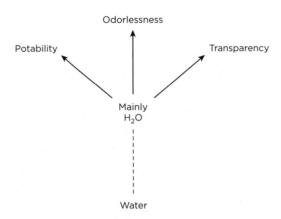

FIGURE 5.4. The essentialist theory of water, after learning about H_2O.

What can the essentialist say in reply? First, they might propose that what goes into the essence position is not "largely made up of H_2O" but some more sophisticated property involving H_2O that swamp water and seawater possess but blood and tea lack. What is that property, then? As the classical concept theorists found, it is extremely difficult to find a definition that picks out all and only the things we count as water (see Section 10.2). The essentialist must succeed in this already formidable task under an additional constraint, that what figures in the definition can be reasonably believed to cause water's characteristic observable properties. I do not think they will succeed.

Second, the essentialist might argue that ordinary thinkers fail to infer, from their rudimentary knowledge of chemistry, that the essence of water is H_2O. Perhaps they fail to notice that "being largely H_2O" is what causes water's observable properties, or perhaps they see it but fail to update their theory of water to reflect their new knowledge. Such a defense is both ad hoc and at odds with the psychological facts: the notion that the essence of water is H_2O is, far from being obscure, widely accepted by chemical novices, philosophers of language, and many other regular human beings.

Third and finally, a defender of essentialism might claim that thinkers' beliefs about the essence of water do not have the inferential significance that is imputed to them by the structure of the essentialist theory, on the grounds perhaps that in the heat of everyday categorization, we rely on heuristics that bypass our beliefs about essence. In that case, you would expect that careful reflection would weaken our conviction that, say, tea is not a kind of water, but this prediction is not borne out.

Let me sum up. The essences in psychological essentialism have two valuable cognitive functions. First, they relieve thinkers of having to represent explicitly the grounds of their basic natural kind categories. That work is outsourced, as it were, to nature (as in Kripke/Putnam accounts of reference). Second, they impart a certain causal structure to the theory that constitutes the concept, a structure that better explains than any other theory of concepts the inferences that ordinary humans make about basic natural kinds: their responses in the Keil transformation studies, their ability to make Kripkean discoveries, and various other inferential dispositions described in Strevens (2000).

The arguments above show that essences are in other ways a liability: they derail any attempt to explain how thinkers incorporate even basic scientific knowledge about water or about the biology of folk genera into their naive understanding of chemistry and biology. Psychological essentialism works best, it seems, when explicit thought about essences is left off the mental table.

Might it be possible to have the advantages of psychological essentialism without attributing a central cognitive role to essences themselves? Is there some way to have the crucial causal structure, and to avoid the need to spell out grounds for category membership, without attributing essentialist commitments to ordinary humans? There is indeed.

CHAPTER 6

Conceptual Inductivism

6.1 Causal Minimalism

Psychological essentialism's explanatory advantage is derived from its attributing to ordinary reasoners what I called in Chapter 5 the core causal beliefs—beliefs such as *There is something about being a swan, namely a swan essence, that causes whiteness*—and its treating such beliefs as ordinary empirical commitments, hence as subject to confirmation or disconfirmation by new information.

If you want essentialism's power to explain but not its essences, then, you need empirical beliefs with much the same causal structure as the core causal beliefs but without the imputation of essential properties. Strevens (2000) suggests simply deleting, from the formulation above, the words "namely a swan essence." The resulting core causal belief (adapting the term to the new, essence-free representations) will then have the following form: *There is something about being a swan that causes whiteness*. Category membership is causally linked by this belief to an observable property, but without implying the existence, at any point in the causal chain, of an intermediary essence.

Combine a set of such beliefs and you have a causal theory with a structure almost, but not quite, identical to that of an essentialist theory. Whereas essentialist theories have the sort of organization shown in Figures 5.2, 5.3, and 5.4, the essence-free organization I am describing might be depicted, for the swan and water categories, as shown in Figure 6.1, with

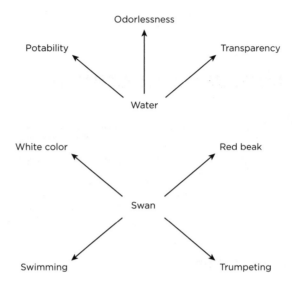

FIGURE 6.1. Causal minimalist theories of water and swanhood.

an arrow between a kind K and an observable property C now interpreted to mean *There is something about Ks that causes C*. I propose that ordinary human basic natural kind concepts have this essence-free structure, a thesis I will call *causal minimalism*.

To what do minimalist core causal beliefs commit the believer? What is it to believe that "something about swans" causes whiteness? It is to believe that there is some internal property P robustly possessed by swans that, by way of a certain causal mechanism, causes whiteness.[1] What is it for swans to "robustly possess" P, then? It is for swanhood to be entangled with P, in the technical sense of "entanglement" defined in Strevens (2008a, 2012b, 2014). A brief and partial characterization of entanglement will suffice here: to say that swanhood is entangled with a causal property P is to

---

[1] As with essentialism, more should be said about the sense in which P is supposed to be "a cause" of whiteness (see note 7 of Chapter 5). I offer the following rough specification: first, P encompasses the internal aspects of some stage or sequence of stages in the causation of whiteness; second, the external conditions required for this stage to have its effect are normally satisfied. As the reasoner becomes more sophisticated, they will replace the second clause with a more useful theory of the nature of the required conditions, as explained in Strevens (2012a).

say that there is a counterfactually robust but not necessarily exceptionless connection between being a swan and possessing P, or in other words, that most or all actual swans have P, that they would continue to have P under a wide range of non-actual circumstances, and that in a wide range of circumstances that would bring non-actual swans into existence (for example, a counterfactual mating between two swans who in the actual world never meet), the swans in question would have P. There will be more on entanglement, as it becomes more important to my story, in Section 12.5. (There you can also find, in Figure 12.1, a pictorial representation of a causal minimalist theory of swans that distinguishes the entanglement and causation relations that are in Figure 6.1 bundled into a single arrow.)

Both essentialism and minimalism hold that, for any putative characteristic property of a basic natural kind, ordinary thinkers posit a hidden cause. For the essentialist, the hidden cause is the essence; for the minimalist it is the property P.

Essentialism and minimalism differ in their psychological consequences in two ways. First, for the essentialist thinker, the hidden causal property is the same for every observable property. A single thing—the swan essence—causes whiteness of plumage, redness of beak, the disposition to trumpet, and so on. For the minimalist, the "something about swans" that causes whiteness may be different from the something that causes redness of beak. Second, for the essentialist thinker, the hidden causal property is necessary and sufficient for category membership. For the minimalist, it need be neither: although there is a robust counterfactual connection between being a swan and having the whiteness-causing property P, some swans may nevertheless lack P, and it may appear in many non-swans (if, say, the causes of whiteness are the same in swans as in certain ducks).

A minimalist's causal theory of swanhood does not rule out the essentialist commitments—it allows that the same property may cause all of a kind's characteristic observable properties, and that this property may be necessary and sufficient for, and indeed the ground of, category membership—but nor does it insist on them.[2] Minimalism therefore attributes to the

---

[2] Essences easily satisfy the criteria for entanglement: if a category has an essence, it will be (tightly!) entangled with that essence.

thinker a set of beliefs that is strictly weaker than those attributed by essentialism: one way to have the minimalist theory structure is to have the essentialist theory structure, but there are many other ways too.

Although weaker than essentialism, causal minimalism in most cases makes identical predictions concerning thinkers' inferences about basic natural kinds. To see why, compare the essentialist theory of swans shown in Figure 5.2 with the minimalist version in Figure 6.1. The sole difference is essentialism's interposition of a swan essence between the characteristic properties and the property of swanhood (that is, the property of belonging to the swan category). Because possession of the essence is necessary and sufficient for swanhood, the inferential transition between them is automatic, indeed compulsory. The conditions under which inferential transitions between observable properties and category membership will be made, then, are in general the same in both an essentialist and a minimalist theory. Simple categorizations based on characteristic observable properties, for example, will tend to follow identical patterns, running from characteristic effects to membership in the category that causally explains those effects (in the absence of other likely causes).

The same goes for more complicated categorization tasks. Minimalism will, for example, explain the results of Keil's transformation experiments just as essentialism does: older children and adults do not consider a transformed raccoon's skunk-like appearances and behavior to be evidence for skunkhood because they see that those properties were not caused by the mechanisms that cause skunk-like appearances and behavior in real skunks.

Likewise, the minimalist and the essentialist account offer more or less identical explanations for the possibility of Kripkean discoveries, by pointing to the core causal beliefs' susceptibility to empirical disconfirmation. Although minimalist core causal beliefs differ slightly from essentialist core causal beliefs, the schwann stories told above go through either way.

The exceptions to the rule that minimalism and essentialism predict identical inferential patterns are those cases where the additional theoretical commitment of the essentialist theory comes into play—that commitment being that the "somethings" responsible for each of a kind's characteristic observable properties are the very same thing, a thing that is also necessary and sufficient for category membership. But this commitment is precisely what is responsible for the shortcomings of essentialism described in

Section 5.4; namely, essentialism's inability to provide a basis for the smooth transition to anti-essentialist notions of species in modern biology and its inability to make sense of Malt's findings about the connections between beliefs about waterhood and beliefs about H_2O.

Minimalism faces no such difficulties. First, the causal minimalist theory structure is compatible with modern biology: a bare minimalist theory need not be amended at all to conform to the past few centuries' revelations, and a minimalist theory that has been augmented with beliefs that have turned out to be false—essentialist presuppositions, in particular—can be revised in the light of these discoveries without its losing its underlying minimalist structure. Second, the minimalist theory structure does not compel chemically literate thinkers to represent some H_2O-related property as being necessary and sufficient for membership in the water category. Malt's data are therefore quite compatible with minimalism. I do not think that minimalism in itself can explain why, say, tea is considered not to be a kind of water—some supplementary psychological principles are required—but while mere consistency with the evidence is a rather modest virtue, it is infinitely preferable to the alternative.

6.2 The Corollary to Minimalism

I have emphasized that minimalism is quite compatible with humans' possessing a range of detailed beliefs about essences. But I hold as a corollary to minimalism that such beliefs are extremely rare. Even experts in chemistry and biology do not, as a rule, represent anything in particular as the essential nature of water or tigerhood.

How, you might wonder, can that be? What is an expert about a category unless it is someone who understands the basis of category membership, that is, someone with accurate beliefs about the category's essential nature? Answer: an expert is someone with a good theory of the category in question, good enough to make reliable categorizations and to support various other relevant inferences. An expert's theory might be, but need not be—and this is my point—metaphysical. After all, are the expert physicists the people with the right metaphysical theory of electrons? No; they have no such theory at all. But they do have a rich enough understanding of the causal and other properties of electrons that they can answer almost any nonphilosophical electron question you might pose. In particular, they can tell you which particles are electrons and which are not, and justify their reasoning to the nth degree—all without relying on a single metaphysical tenet.

In the same way, an arboreal specialist with rich enough core causal beliefs about, say, California nutmegs—an expert with intimate knowledge of the appearance, behavior, and internal structure of California nutmegs, all represented as causally connected to the category in the minimalist manner on display in Figure 6.1—will be able to reliably identify these trees wherever they are found, and to satisfy the most persistent questioner. Or at least, they can afford satisfaction provided that they are not asked about the metaphysics of California nutmegs, concerning which they have no opinion, and need none.

What about philosophers? Some have published views about the essential natures of substances or species; surely they have (as minimalism allows) corresponding beliefs? Indeed they do, but even in these individuals, I suggest, beliefs about essences do not play a full-blooded role in cognition: although they figure, of course, in conscious and deliberate reasoning in which they are invoked explicitly, they otherwise have little or no impact on category-related judgments. They are, in particular, sidelined in day-to-day case judgments: whatever goes into a philosophical essentialist's everyday decisions about whether some animal counts as a tiger or a dog or a swan, it is not their professional metaphysical commitments.

Similarly, I will later propose, even a philosopher with a considered view of the nature of knowledge makes everyday judgments about whether some belief counts as knowledge that circumvent their own thesis (except of course when they explicitly refer to that thesis in conscious and deliberate reasoning). There is a Rawlsian veil of a sort at work here, but its function is the mirror image of its function in conceptual analysis, modern or hypothetical: rather than shielding the definition or essence postulate at the center of the mind's quotidian case judgments from conscious scrutiny, it shields the case judgments from any consciously entertained definitions or essence postulates.[3]

---

[3] Nagel (2012, 501) makes a similar point.

Could the impotence of explicit theory in everyday judgments be explained by the "quick and dirty" or untheoretical nature of the judging process? Not if the minimalist theory of natural kind concepts is correct: minimalism says that everyday judgments are in the relevant sense often highly theoretical, resulting as they do from the reasoned application of the logic of causality to the core causal beliefs.

Some other explanation of the reverse veil is called for. My guess is that it is more psychologically than philosophically interesting, so I will not answer the call (though it is surely connected to ordinary categorizations' "type 1½" ability—see Section 4.3—to "hide in plain sight").

If my corollary is correct, then, although minimalism permits beliefs about essences, such beliefs are rare, and even where they do occur they are typically cognitively inefficacious. It seems that we humans are eager to take advantage of minimalism's invitation to avoid mental metaphysics.

⸺

Causal minimalism is not causal nihilism; it builds less into the theories underlying basic natural kind concepts than does psychological essentialism, but like essentialism it attributes to those theories a "starburst" structure. As far as we know, nature on the whole respects that structure, but in individual cases, categories have been found to lack it. A simple and classic example is jade, which turns out to be either of two distinct substances, jadeite and nephrite. Another class of examples are the polyphyletic groupings in biology, such as the moles, which unite discrete segments of the evolutionary tree, often with respect to resemblances due to convergent evolution.[4]

Suppose that the original naive concept of jade had, as causal minimalism implies, the starburst structure (imagining for the sake of the argument, then, that humans originally regarded it as a basic natural kind like water or gold). That structure attributes to jade a single underlying property responsible for, say, its hardness. There is no such property: what explains the hardness of nephrite is not what explains the hardness of jadeite.

---

[4] I understand the moles *sensu lato* to include the "true moles," the golden moles, and the marsupial moles. In a narrower sense the word refers only to the true moles, which are closer to a natural kind (technically, a paraphyletic group), illustrating the centripetal effect of objective taxonomical structure (see Sections 10.4 and 12.3).

Upon learning this fact, how did we manage to maintain the concept of jade with its original extension? Rather than identifying jade with either nephrite or jadeite or concluding that there is actually no such thing as jade?

The answer must be that the starburst structure is not in fact, for individual concepts, psychologically fixed. Ordinary learning can transform our naive theory of jade or moles from one that has the starburst structure to one that does not. That indeed seems to be not merely consistent with but a consequence of my proposal that the core causal beliefs are ordinary empirical beliefs like any others, ever capable of revision or refutation. There is nothing we could not, given the right sort of evidence, come to believe about jade or moles, and likewise, no belief about jade or moles that we might not, given the right sort of evidence, come to abandon.

What is the force of causal minimalism, then? First, it explains the characteristic patterns of human judgment about basic natural kinds, such as our tendency in Keil's transformation studies to regard a cosmetically altered raccoon, in spite of its skunk-like appearance, as a raccoon. If our beliefs about a kind were so profoundly revised in the light of empirical evidence that the corresponding concept's cognitive structure entirely lost its starburst structure, we would no longer make the Keil inference (or if we by chance did, it would be for some unrelated reason).

Second, it explains what we are inclined to count as a natural kind. After making our discoveries about jade and moles, we no longer regard those categories as full-blown natural kinds, because we do not consider them to have the correct causal structure. Psychological essentialism makes the same prediction, but it also looks to predict that on discovering the falsehood of biological essentialism, we should cease to count any of our biological taxa as natural kinds—contrary to the facts.

I am, in any case, about to move on from causal minimalism to a more general thesis about concepts, encompassing even the jade and mole concepts, that will provide the foundation for my account of philosophical knowledge. By way of preparation, let me summarize the picture of the mind's natural kind concepts offered by minimalism and its corollary, and pose three profound questions about the minimalist's explanatory strategy.

Every natural kind concept, according to the minimalist, is accompanied by a causal theory spelling out some of the (putative) causal consequences of belonging to the kind. It is the core causal beliefs constituting this theory that mentally connect kind membership to a kind's (putative) characteristic observable properties. Thus it is these beliefs, deployed using the familiar rules of causal reasoning, that are responsible for much of our everyday reasoning about kinds, in particular, the predictions we make about a specimen's properties based on kind membership (*that's a swan, so it will probably have a red beak*), and the case judgment we make about a specimen based on its properties (*given its color, size, shape, and behavior, that bird is surely a swan*).

What minimalism does not posit, and what according to its corollary is normally absent, is any sort of explicit definition, reference-fixer, or essence postulate, or any representation or inferential tendency that tacitly constitutes a psychological commitment to such a thing. We do not define swanhood; we do not harbor any referential intentions for our concept of swanhood; and we do not normally have even tentative beliefs about the nature of swanhood. Nor do we reason as though definitions or referential intentions exist: in our thinking, we take our minimalist theories at face value, treating them as ordinary empirical theories about the corresponding categories.

Our swan concept has its role in thought, then, entirely in virtue of its being embedded in ordinary, corrigible, empirical beliefs about the causal consequences of swanhood. Minimalism makes no appeal to semantic or to metaphysical representations, tacit or explicit; even if they exist, as they do in the mind of the working metaphysician, they are, on the minimalist view, almost always cognitively impotent.

Now the three questions about the minimalist approach.

First, if there is no definition, no essence postulate, no commitment in the great majority of minimalist minds to any particular structure for the categories concerning which they are doing their thinking, what gives thought about those categories—the folk genera, the chemical substances—the stability that it palpably has? There are, I think, two kinds of stability to explain. One is the actual stability of our thought's subject matter: as our theories improve, we hold changing opinions about fixed categories. To put it another way, the extensions of our concepts are typically (though not always) stable. It is the nature of reference, with an assist from the world, that underwrites this feature of thought; in Chapter 9 I will sketch a theory of reference for minimalist concepts in which extensions tend to remain stationary even as beliefs evolve. The other kind of stability is the internal perception of extensional stability: it seems to us that the extensions of our concepts change only under exceptional circumstances (as when "corn" contracted to include only maize). This seeming is, I would guess, a complex phenomenon, resulting from a mix of experience, the short-term continuity of thought (in which the synonymy of different instances of the same conceptual token structure within a single inferential step is tacitly presumed), and perhaps some "folk semantics." In any case, I will put the question of internal stability aside for the duration.

Second, there is a forceful argument, based on Fodor's (1981) argument for concept nativism, that if our beliefs about folk genera and chemical substances were entirely empirical, we would be unable to acquire novel genus or substance concepts. The problem of acquisition is posed and solved in Section 15.1.

Third and finally, even if most of us do not have definite beliefs about water's essential nature, we do seem to have—after some philosophical tutelage, at any rate—beliefs about necessary conditions for waterhood. It is plausible, for example, that in order to count as water, a substance *must be* mostly (or if not mostly, then partly) made up of H_2O. Theories of natural kind concepts that posit definitions or reference-fixers can easily explain such convictions. A minimalist explanation of the same is given in Section 9.6.

6.3 Inductive Concepts

From concepts of natural kinds to concepts of the philosophical categories—that is the leap I want to make. I do not propose that causal minimalism with its starburst structure and its notion of a core causal belief is sufficient to characterize the concept of, say, knowledge; that is hardly likely. Causal minimalism can, however, be seen as an instance of a more general thesis about the nature of concepts that might not unreasonably be applied to philosophical concepts.

This thesis I call *conceptual inductivism*. Inductivism concerning a concept begins with the claim that the concept's cognitive significance is exhausted by the ordinary or empirical beliefs in which it figures. It therefore denies that the concept's cognitive significance derives in any way from a definition or referential intention, whether explicit or tacit, or a belief about the corresponding category's essential nature (an essence postulate). Even where such definitions, intentions, or essence postulates clearly exist—in philosophical dogma or conversational asides—inductivism denies them a significant effect on judgments of category membership. (As you can see, inductivism is more a generalization of minimalism plus its corollary than of minimalism alone, since bare minimalism is strictly speaking compatible with any amount of defining, intending, and metaphysical musing.)

The central inductivist notion is that of an *ordinary belief*, a belief about a category that is not regarded by or otherwise treated by the believer either as articulating the category's essential nature or as holding true in virtue of its logical or conceptual connections to the category's essential nature (for example, in virtue of its being a logical consequence of the essential nature). What makes a belief ordinary, then, is the lack of a certain tacit or explicit attitude. The believer might take the contrary attitude—they might regard their belief as certainly not logically or conceptually connected in any way to the relevant category's underlying nature—but equally, they might simply not have an opinion on the issue. Such open-mindedness, whether mindful or not, is quite sufficient for ordinariness, that is, for ordinary beliefs' having an inferential role in the mental economy that exactly reflects their ostensible content.

There is more to inductivism than pervasive ordinariness, but it takes the form of a research strategy rather than a psychological hypothesis: the inductivist, wherever possible, attempts to explain central features of a concept's cognitive significance by appealing to ordinary beliefs about explanatory connections. Causal minimalism's core causal beliefs are a prime example: they connect membership of a concept's corresponding category—the basic natural kind in question—to other observationally more accessible properties by way of a causal-explanatory link.

When the possessor of an inductive concept reasons about the corresponding category, then, their thinking tends to take the form of explanatory inference, either going from explanandum to explainer (broadly, inference to the best explanation), or from explainer to explanandum (e.g., from cause to effect). Not all reasoning, of course, will follow explanatory links. Good inferences may take advantage of correlations, patterns, and other connections with no discernible explanatory significance; equally, they may go by way of the laws of deductive logic. Inductivist psychologists strive, nevertheless, to explain interesting inferential behavior by positing mental representations of explanatory relations connecting category membership with other features of the world.

It will help to crystallize some of these ideas, I think, if I explain why I use the term "inductivism." A description of a category's essential nature is, if correct, a basis for apodictic categorization judgments: it provides an infallible, because constitutive, criterion for category membership. An ordinary true belief typically does not confer this sort of infallibility. Even if I have a complete and correct theory of the causal consequences of swanhood, and I apply it without making logical errors, I might nevertheless judge falsely that, say, a molecular reproduction of a swan is a swan (see Sections 10.3 and 10.7). The source of this fallibility is the nature of the ordinary beliefs: they tend to articulate explanatory links between category membership and the properties on which a categorization is based, such as observable appearances and behavior. Reasoning from the categorization basis to category membership is therefore an inductive process that like all inductive reasoning may in principle go astray no matter how expertly it is conducted. The view of concepts proposed here implies, then, that categorization and other category-involving inferences will typically and persistently fall short of the deductive, certainty-transmitting character imputed to categorization by the classical theory of concepts and its offspring. It is to remind you of this provisional mood that I call the view conceptual inductivism.

A further source of the "inductive feel" of inductivist concepts is that the ordinary beliefs that make up an inductive concept's cognitive structure are typically not knowable with certainty. Even should we achieve a complete, correct theory of swans, we will not know it as such. At the end of science, when all the empirical facts are in, there will still exist some in-principle uncertainty as to whether we have interpreted the evidence correctly. That said, although in the vast majority of cases we lack complete certainty about our categorizations (even given certainty about the categorization basis), we are often highly confident in our conclusions. This confidence will play a role in my explanation of "case certainty" concerning philosophical case judgments in Chapter 14.

The term "inductivism" is allusive rather than exact. Some categorizations using an inductive concept are deductive, as when I classify a swan as not being an item of knowledge because it is not a mental state. Some ordinary beliefs (those of certain philosophers, for example) represent essential natures, and might even be treated by the believers as infallible. As a name, then, "inductivism" is supposed to capture the general flavor of the view and its consequences rather than to lay down an exceptionless epistemological or psychological law.

Another quite orthogonal question about my term "conceptual inductivism": why not call the theory of concepts so named simply the "theory-theory," the view that concepts are mental theories (Section 5.3)? After all, the view that concepts get their cognitive significance from their featuring in ordinary theoretical beliefs is arguably, far from being a newfangled idea, simply the theory-theory of concepts in its purest, clearest, barest version.

There are, nevertheless, three reasons to avoid the "theory-theory" name. First, it is lexicographically confusing, even faintly ridiculous.

Second, it suggests that the ordinary beliefs to which a concept owes its significance—the ordinary beliefs that constitute its cognitive structure—must form a coherent explanatory body of propositions worthy of the name "theory." While inductivism does indeed seek systematic explanatory beliefs wherever they can be found, it is quite ready to acknowledge that many concepts may owe their cognitive significance to sets of beliefs that are ramshackle and shallow. Nor will the inductivist insist that the distinction between the theoretical and the ramshackle is anything other than one of degree, or attribute any special importance to qualitative distinctions among bodies of belief of different degrees of systematicity and intrinsic explanatoriness even if they should turn out to exist. In short, the first "theory" in "theory-theory" is potentially misleading as well as vulnerable to misreading.

Third—and this is no fault of the psychologists who have proposed and developed the theory-theory—it has had foisted upon it by philosophers a commitment that is quite contrary to the spirit of inductivism. Here, in Fodor (1998, 117), is the sort of hermeneutic loading I want to avoid:

> Medin and Wattenmaker (1987; see also Murphy and Medin 1985) undertake to "review evidence that suggests concepts should be viewed as embedded in theories" (34–5), a thesis which they clearly regard as tendentious, but which, as it is stated, it's hard to imagine that anyone could disagree with. What I suppose they must have in mind is that concepts are somehow *constituted* (their identity is somehow determined) by the theories in which they are embedded. But that claim, though tendentious enough, doesn't amount to a new account of conceptual content; unless the 'somehows' are somehow cashed, it just reiterates [inferential role semantics].

One part of Fodor's unwelcome baggage is the supposition that the theory-theory is principally an account of concept individuation, counting two concepts as distinct just in case they play substantially different theoretical roles. That, I think, mischaracterizes the explanatory use to which cognitive psychologists put the theory-theory; they are for the most part, as Fodor himself remarks in the preamble to this passage, quite uninterested in concept individuation. (Even those psychologists who study conceptual change have no need of a criterion for individuating individual concepts, I explained in Section 4.1; they require nothing more than a taxonomy of concept types.)

The burden on the theory-theory becomes even more unwholesome when concept individuation is assumed to respect the principle that concepts are to be individuated by their content, because it then follows that conceptual contents are to be individuated in the same way that theories should be individuated (whatever way that is), which leads directly as Fodor observes to some sort of inferential role semantics. Such a semantic theory is technically consistent with conceptual inductivism (though as explained in Section 7.4 some of its close cousins are not). But it suggests, even if it does not insist on, a semantically driven strategy for understanding cognition that is quite at odds with the inductivist strategy driven by ordinary explanatory beliefs.

Inductivism ought to be framed in a way that makes plain its complete disregard for meanings, conceptual truths, definitions, and any other facets of semantics above and beyond those involved in ordinary inductive reasoning, such as the logic of probability, causation, and inference to the best explanation. My final reason to abandon the "theory-theory" label is to shake off all lingering connotations of inferential role semantics.

I will say, to avoid the air of paradox, that like everyone I assume that concepts acquire their semantics in virtue of their inferential roles. (With the demise of the picture theory of mental representation, what else do they have going for them?) That is, indeed, a consequence of the dispositional approach to reference advocated (independently of inductivism) in Chapter 9. The dispositional theory will not, however, individuate content according to inferential role; rather, it will give the same intension to concepts with quite different inferential roles, and in some cases different intensions to concepts with identical inferential roles. More generally, it will take an extensional rather than an internal approach to understanding semantic significance—an approach in which the function of inferential roles is to get concepts attached to the world rather than to serve as a self-standing repository of semantic value.

Inductivism, then, is a version of the "theory-theory" of concepts that as far as possible abandons all Fregean baggage. In place of semantics, it invokes the thinker's representation of explanatory structure—the starburst structure of causal minimalism being one among many possibilities—to explain why particular classes of concepts, such as the basic natural kind concepts, have their distinctive cognitive character.

6.4 Generalized Conceptual Inductivism and the Myth of Depth

Almost all concepts, I propose, are inductive. Concepts of individuals are inductive. Concepts of events are inductive. Concepts of actions and artifacts and utterances are inductive. Most important, the concepts of the philosophically interesting categories are inductive. That is the thesis of conceptual inductivism generalized. If I am correct, most concepts in most heads have the cognitive significance that they do in virtue of their figuring in merely ordinary beliefs. The typical concept, then, has no definition and is associated with no referential intention or other reference-fixer. Nor does its possessor harbor any particular beliefs or even speculations about the essential nature of the corresponding category, that is, about what makes things members of the category. Thinkers simply go about their cognitive work using their beliefs about categories to make judgments both of category membership and of the consequences of category membership, in the same way that owners of minimalist natural kind concepts use their core causal beliefs to do the same.

That the great majority of concepts are inductive is a thesis to be distinguished from Quine's denial of a distinction between analytic and synthetic propositions. For Quine, when a belief changes in response to a new observation, there is no fact of the matter whether it is being revised like a definition so as to more economically organize the inventory of accumulated evidence, or undercut like an ordinary belief, that is, refuted or disconfirmed by the freshly observed fact. All change in mental representation, according to Quine, inhabits a gray zone somewhere between meaning change and change in our representation of the external world. My view is quite different: semantic revision and empirical (or other) refutation can be distinguished, and a change to an ordinary belief is purely the latter. For Quine, there is no fact to ground the distinction between a stipulation and a firmly held ordinary belief; for me, the distinction is unproblematic, but almost all our mental representations are ordinary beliefs rather than stipulations.

Are there any mental stipulations at all? Yes—almost surely, some concepts are not inductive. The clearest cases are concepts created by artful definition, such as the concept of topological compactness in mathematics (a set is compact just in case all of its covers have a finite sub-cover). Quite plausibly, the same is true of the concept of a prime number. Again, I am not a Quinean; I allow that concepts can be created by definition, and I am happy to acknowledge that the possibility is in some cases actualized.

All the same, I hold that there is a natural psychological tendency for concepts to come into being unattached to definitional or metaphysical beliefs and to remain ordinary thereafter. The tendency can be overridden, but typically it is not. Thus, the great majority of our concepts are inductive, including our concepts of most or all of the important philosophical categories—not categories corresponding to philosophical terms of art, but categories knowledge of the nature of which constitutes a philosophical end in itself, such as beauty and justification, causality and the good.

⁓

Conceptual inductivism is in many ways a simple, obvious view. What could be more sensible and workmanlike, more honest and democratic, than the thesis that our beliefs are overwhelmingly "ordinary," that they function in cognition in just the way you would expect, if you were to take them at face value? Yet I have barely seen it defended in three millennia of philosophy and psychology.

The reason, I conjecture, is that our theorizing about concepts is profoundly in the thrall of what I call the "myth of depth." According to the myth, the reliable classification of things into "deep" categories such as natural or philosophical kinds requires "deep" principles, which is to say, principles that articulate the ultimate basis of category membership—that spell out a category's essential nature. This goes along with a tendency to think that a concept of such a category cannot be possessed without grasping to some extent the basis of category membership, that is, without building a definition or at least committing to an essence postulate for the kind in question.

It is the myth of depth that has inclined many philosophical and psychological thinkers to suppose that concepts are built on definitions, that representations of essential natures can be found in the head, and that expertise in a domain is in part a matter of knowing the ultimate criteria for membership in the domain's categories. The concomitants of the myth are many:

> A mature science must define its terms.
> To represent causal connections between things or membership in basic natural kinds, a thinker must have an incipient metaphysics of causality or natural kindhood. Young children and other small mammals must, then, for want of the requisite philosophical sophistication, conceptualize the world in terms of appearances instead of deep categorization schemes.
> A profound change in scientific worldview must be accompanied by a change in the concepts of causation and explanation.

Conceptual inductivism undercuts these apparent necessities. It is possible, in the case of natural kinds, to make categorizations that are reliable, well founded, even "deep" in the sense of being sunk into the explanatory bedrock of the world, without having any opinion whatsoever about the nature of the kind in question. In the case of biological taxa and chemical substances, the experts at determining category membership know, and care to know, almost nothing about the metaphysics of kinds. Thinkers with inductivist concepts skate over the surface of things from a philosophical point of view, yet their ability to think sophisticated thoughts about both appearance and reality does not at all suffer as a consequence. The myth, then, is just that.

6.5 Are Philosophical Concepts Inductive?

There is a strong case to be made that our concepts of chemical substances and folk genera are inductive. But why think that philosophical concepts are inductive? Who but a poet would compare knowledge to water?

I would. Just as the water concept is given cognitive significance by a set of ordinary beliefs or hypotheses about water, I suggest, our concept of knowledge may be given cognitive significance by a set of ordinary beliefs about knowledge—beliefs about features of knowledge that are not represented or otherwise treated as true by definition or on any other semantic grounds, that are not by their very nature immune to disconfirmation by new information, and that do not purport to describe the essential nature of knowledge. In the same way that the beliefs comprising our water concept make claims about the typical properties of water without asserting that these properties are somehow constitutive of waterhood, so the beliefs comprising our knowledge concept would make claims about the typical properties of knowledge without asserting that these properties are constitutive of knowledge. If the claims are rich enough, such a theory can serve as a powerful tool for judging what is knowledge and what is not, without pretending in any way to provide a philosophical theory of knowledge, an analysis of what knowledge really is—of what ultimately makes one thing an item of knowledge and another thing not.

6.5. ARE PHILOSOPHICAL CONCEPTS INDUCTIVE?

As to what sort of claims might constitute an ordinary theory of knowledge, one answer is that the theory's central propositions are none other than the various claims about knowledge that epistemologists have attempted to transform into philosophical analyses of knowledge. They would include perhaps the following:

Knowledge is a certain kind of justified true belief.
Knowledge is a mental state produced by a reliable belief-forming process.
A body of knowledge will tend to form a maximally coherent whole.

Considered as nothing more than ordinary beliefs, these claims are quite consistent with one another (at least if they are understood as generic claims that allow exceptions); it is only when interpreted as philosophical analyses spelling out the essential nature of knowledge that they are mutually exclusive.

Another answer (perhaps expanding upon rather than replacing the first) is even more in the inductivist spirit. It proposes that the central propositions of the ordinary theory of knowledge are hypotheses that assert a place for knowledge in an explanatory network comprising belief, justification, and other epistemic states and relations. This possibility will be further explored in Chapter 13.

Either way, inductivism suggests that, when supplemented with equally ordinary theories of justification, belief, and so on, the ordinary theory of knowledge will be sufficiently powerful to classify, by way of inductive inference, almost any specimen as either knowledge or non-knowledge. In virtue of our subscribing to such a theory, we will be experts about knowledge, without knowing what it is.

Perhaps you are skeptical, nevertheless, that the rich categorizing power of our philosophical concepts could inhere in a set of perfectly ordinary beliefs, as opposed to a belief about the corresponding category's essential nature. Let me give you a reason to reconsider.

Suppose that the concept of knowledge has at its core a representation, either putative or stipulative, of the true nature of knowledge. Suppose further that this representation takes the form of a proposition. Such a proposition can be divided into two parts: an ordinary claim, and a further claim that the ordinary claim in fact represents the true nature of the category in question. For example, the proposition stating that what it is to be knowledge is to be justified true belief can be divided as follows: (a) the ordinary proposition that all knowledge is justified true belief and vice versa, and (b) the proposition that (a) captures the true nature of knowledge. The second part of the criterion can be subtracted without disturbing the content of the first (trick cases aside). Thus for every possible proposition describing the essential nature of a philosophical category there is a corresponding ordinary theory about that category that has much the same content.

The metaphysical proposition and its ordinary counterpart are fraternal, not identical, twins; their cognitive significance differs in subtle ways. But the ordinary theory surely has a richness and depth in its categorizing power that is comparable to, if not exactly equal to, the metaphysical proposition's categorizing power. It cannot be maintained, then, that propositions about essential natures in general provide far superior grounds for categorization than ordinary theories. Indeed, for any putative representation of an essential nature there is an ordinary counterpart that is so close in its categorical implications that, if you take the former as a plausible candidate for the psychological core of the concept of some philosophical category, you must consider the latter to be a plausible candidate as well. Thus, if you think that it is reasonable to suppose that all philosophical concepts have beliefs about essential natures at their cores, then you are also committed to supposing that some or all of them may have perfectly ordinary theories at their cores, and therefore that inductivism about the philosophical concepts is a real and robust theoretical possibility.

I will, in any case, baldly presume in what follows that generalized conceptual inductivism extends to our concepts of philosophical categories. The case for inductivism is hardly conclusive; it is an exciting and largely unexplored working hypothesis rather than an empirically established truth. But in this book, at least, it will be questioned no further.

CHAPTER 7

Inductivism versus Conceptual Analysis

If philosophical concepts are inductive as I suppose, then many existing accounts of philosophical analysis—including every attempt to understand armchair philosophy as some sort of conceptual analysis—must be mistaken.

The case against modern conceptual analysis, according to which philosophical analysis turns on mental definitions, is straightforward. Inductive concepts have no definitions. A self-conscious analyst searching for a definition will therefore fail. For the same reason, if ordinary working analysts succeed in doing something of value, it cannot be for the reason given by the modern analysts. Definitions can play no part in philosophical methodology, because as philosophers and psychologists have for decades now suspected, there are none.

Hypothetical conceptual analysis aims not to uncover a mentally buried definition but rather a metaphysical belief, or what I have called an essence postulate. To undertake the hypothetical analysis of, say, your swan concept is to attempt to write down your folk theory of the nature of swanhood, of what it is that makes something a swan. That enterprise, too, must fail: either there is no such theory, because like most people you have no essence postulates, no beliefs about the nature of swanhood, or—if you moonlight as a biological metaphysician—your only such beliefs are the philosophical theories that you consciously entertain, and thus of which you were already fully aware before analysis began.

What about accounts of philosophical analysis that, like modern analysis, turn on the stipulative status of certain conceptual content, but which avoid positing that such content takes the form of an explicit definition? They too must fail, because an inductive conceptual psychology is inconsistent with any sort of stipulativity, direct or indirect, explicit or tacit. Much of what follows seeks to illustrate this claim, using a series of case studies that, even if they do not quite add up to a deductive argument for the general incompatibility of conceptual inductivism and stipulativity, gesture emphatically in that direction.

7.1 Intensional Analysis

Intensional analysis eschews definitions and (when practiced self-consciously) aims to infer knowledge of a category's essential nature from knowledge of a stipulative reference-fixer.

As in Chapter 3, let me begin for expository reasons by discussing the exceptionally simple account of reference-fixing on which new concepts are introduced by explicitly bestowing upon them a reference-fixing rule. When first encountering specimens of gold, for example, I say to myself—waving at a stack of doubloons and coining a new word—"Let 'gold' refer to any metal having roughly the same microstructure as these specimens." Because the rule is given by stipulation, it is guaranteed to be correct and, by the same constructivist line of thought that characterizes modern analysis, to capture the essential nature of the category picked out by "gold"—to capture, that is, the fundamental reason that a substance counts as gold.

Some sort of mental veil (it is supposed) obscures the stipulated reference-fixing rule, but it can be recovered by the method of cases, that is, by making judgments about particular specimens (typically in hypothetical scenarios where the relevant physical facts are taken as given). Because such case judgments are regulated by the reference-fixing rule, they function as evidence for or against various hypotheses about the content of the rule, and so can be used to discover its true form, and thus the corresponding category's essential nature. And that is how philosophical analysis provides philosophical knowledge.

If the concept of gold is inductive—if, for example, it is as the causal minimalist says—then none of this can be correct. Were "Gold is metal" to be stipulative, then it would be incorrigible: neither empirical evidence nor further reflection could provide you with reason to regard it as false. (You might decide to "re-stipulate," associating "gold" with a new reference-fixing rule, but that is quite different from concluding that the old stipulation is factually incorrect.) Were "Gold is metal" to be, by contrast, an ordinary belief, as inductivism requires, then it could be refuted (or supported) in all the usual ways: chemical assays, computational modeling, testimony, and so on. Either the concept's stipulative status or its inductive nature has to go. The psychological evidence suggests we should ditch the former, in which case intensional analysis is left without any foundation.

Before moving on to more sophisticated varieties of intensional analysis, let me consider two replies on behalf of the simple version that are of quite general interest.

An empirical psychological theory such as causal minimalism concerns the inferential tendencies of ordinary reasoners. Might it be that, although basic natural kind and other inductive concepts are associated with reference-fixing stipulations, their users typically disregard these stipulations and deploy the terms in thought in accordance with the inductivist theory? In making this suggestion, the intensional analyst does not renounce the reference-fixing power of the stipulation; that is central to their conception of philosophical analysis. Rather, they posit that thought goes on largely in ignorance of, or at least ignoring, this reference-fixing power. The stipulative nature of the reference-fixing rule has, in short, no impact on the cogitations of ordinary, unreflective thinkers, including working philosophical analysts. Such individuals reason just as conceptual inductivism implies. Although they are irrevocably conceptually committed, then, to gold's being metallic, they are unaware of the commitment, and so treat the proposition as empirically refutable—a subtle logical incoherence.[1]

The incoherence is, it must be conceded, curable. Careful attention to the workings of language and the mind can transform the ordinary analyst

---

[1] Modern conceptual analysis might be defended in the same way: definitions determine the semantics of our concepts, but we fail to grasp this fact and so we reason inductively in ways that are incompatible with their stipulative nature.

into a self-conscious analyst, who understands that their case judgments (or some subset thereof) are based on a reference-fixing rule made incorrigible by stipulation. It is these thinkers who understand the true nature of philosophical analysis, and by understanding it, provide its vindication.

In positing such a bifurcation, however, the intentional analyst loses much of the modern conceptual analyst's attractive explanation of case certainty. (Here we revisit a form of argument that wound through Chapters 2 and 3.) Self-conscious analysts grasp the stipulativity of certain aspects of their thought, and so grasp the security of certain of their case judgments. But the case certainty experienced by ordinary working analysts is left unexplained, as is the case certainty experienced by stipulativity skeptics such as myself. Why am I so sure about my Gettier judgments? Not, I can assure you, because I have come to see that they are based on a stipulative reference-fixing rule.

Pushing the argument a little further, I suggest that even self-conscious believers in reference-fixing rules exhibit, under the hood, the patterns of inference predicted by conceptual inductivism. They are largely incapable of following through on their own supposed stipulative commitments, because as inductivism asserts, those commitments have no actual purchase in the mind.

The second defense of intensional analysis concedes that the stipulation of reference-fixers will create beliefs about the corresponding category that are in principle irrefutable—a state of affairs at odds with conceptual inductivism—but holds that these beliefs are too weak or insignificant to put any significant constraints on ordinary thought. Conceptual inductivism is, in that case, not quite true, but very close to empirically adequate.

Suppose, as above, that I fix the reference of "gold" by stipulating: "Gold is anything having the same microstructure as this stuff" (pointing at a nearby ingot). Then it is a stipulative truth that the ingot in question is made of gold, but this is hardly likely to have much impact on my future thinking; almost every belief I go on to form concerning gold—that it is yellow, conductive, has atomic number 79, is valuable, and so on—will be deductively unconstrained by my stipulation, and will therefore function as an ordinary empirical belief, just as inductivism demands.

There is one exception: the belief that all gold has the same microstructure is stipulatively protected from refutation by this act of reference-fixing.

It seems to me, however, that the belief is as ordinary as the others—early chemists might quite coherently have questioned whether it was so, in the same way that biologists not only coherently but (in the view of many) correctly questioned whether various folk genera were built around essential properties.

In any case, as many philosophers have remarked, gold cannot be picked out simply by pointing. Your semantically naked extended finger might just as well indicate the grime on the ingot's surface or the air in between. It seems that you need to add some descriptors to zero in on the substance you intend to name—"I mean the metal that makes up the ingot, not whatever is sitting on the surface layer." Further descriptors are required to specify the taxonomic level of the term that is being defined: what's meant is gold, not "24 karat gold" or "noble metal" (Devitt and Sterelny's [1987] "qua" problem). But every such descriptor will add to the stock of stipulative truths.

An act of ostension needs more help still when the term whose reference is to be fixed is theoretical: "proton," or "gravity," or "money." In the case of philosophical terms such as "knowledge" and "causality," pointing hardly helps at all; most of the work will have to be done by descriptors, which inevitably bring with them an entourage of substantial stipulative truths.

I could continue in this vein for some time, considering for example the possibility of fallible descriptors—expressions that help to fix reference even if they are not literally true. (These are discussed, and the stipulative truths that they too inevitably create inventoried, in Strevens [2012c].) But it is hardly necessary, because the strategy of defending intensional analysis by arguing that stipulative reference-fixing creates only weak stipulative truths is self-defeating. The philosophical fruits of intensional analysis are facts about essential natures that are determined by stipulative reference-fixing. If the facts so determined are logically extremely weak, then the facts uncovered by analysis must be at least as weak. Philosophical analysis may be possible, in this view, but it is far from worthwhile. I am looking for a more robust defense of armchair philosophy than that.

A generalized version of this argument, providing what you might think of as a master argument against modern and intensional conceptual analysis, runs loosely as follows: Conceptual inductivism implies that all (or almost all) of our substantive beliefs about categories are ordinary beliefs, subject to refutation in the usual ways. Modern analysis and allied techniques can provide access only to facts that are entailed by stipulative beliefs, which are not ordinary. If inductivism is correct, then, there is little or nothing of substance that these methods could possibly learn.

7.2 Beyond Explicit Stipulation

I supposed, in the previous section, that reference-fixers come into existence by way of explicit stipulations. The mind, in effect, makes a declaration as to the extension of a concept, tokening a mental sentence of the form "Let 'gold' pick out the stuff that ___," and then in its subsequent reasoning honors (or is obliged to honor) the logical implications of the declaration. But as I explained in Section 3.2, there are more subtle forms of "stipulativity" that also have the capacity to fix reference and, in so doing, to determine aspects of a category's essential nature in a way that is accessible to the method of cases.

In a case of tacit reference-fixing, for example, a thinker does not explicitly represent a stipulation of the form "Let 'gold' pick out the stuff that ___," but they act as though they do—their reasoning conforms, at least roughly, to the patterns of reasoning that would be found in a thinker who has made such a commitment and succeeds in doing it justice. They reason, in other words, as though a substance may count as gold if and only if it satisfies the conditions ___ that complete the reference-fixing rule. If certain Quinean objections can be overcome, such inferential dispositions or behavior alone might be regarded as conferring stipulativity on the rule. Could philosophical analysis hinge upon tacit reference-fixing stipulations of this sort?

At the heart of conceptual inductivism is the proposition that "stipulative" inferential behavior is nowhere, in the human mind, to be found: all (or almost all) is corrigibility. Even if, contra Quine, the philosophy of tacit stipulation is sound—even if a tenacious refusal to abandon the belief that gold is metallic in the light of any amount of reflection or empirical evidence does indeed constitute something semantically equivalent to a stipulation—the psychology will not harbor such a thing, which is to say, human thinkers will not exhibit any such tendencies.

This line of argument shows, I hope, that the thesis of conceptual inductivism is not only about little sentences in the brain: it is about any inferential tendency, any feature of our psychology, that plays a belief-like role in steering inference. Inductivism concerning a concept is the view that all such features play the inferential role of ordinary beliefs, not of definitions, presumptive reference-fixers, or metaphysical hypotheses. Or in other perhaps slightly exaggerated words, inductivism means that a concept is in principle infinitely inferentially mutable.

The simple tacit approach to stipulative reference-fixing clashes head-on with inductivism because it requires certain inferential patterns, which are precisely what inductivism rules out. Might a more indirect conception of stipulativity do better?

Suppose, to use the reference-first strategy introduced in Chapter 3, you stipulate (explicitly or tacitly) a theory of reference according to which your basic substance terms, such as "gold" and "water," refer to whatever satisfies your beliefs about the corresponding categories' chemical properties. If you believe that gold is yellow, metallic, and malleable, then, "gold" will refer to whatever substances have those properties.

Clearly, such stipulations create necessary, thus irrefutable, truths in much the same way as direct stipulations about the chemical kind terms themselves: if the reference of "gold" works this way, then gold must be yellow. Provided that this necessity makes itself felt in case judgments, you can use philosophical analysis to learn of the yellowness of gold in the armchair. But by the same token, your belief that gold is yellow is not an ordinary empirical belief, subject to refutation by chemical inquiry. Thus the gold concept is not, on this view, an inductive concept. All of the psychological evidence points, however, to the conclusion that it is in fact inductive, that we can happily and coherently revise our belief that gold is yellow in the same way that early modern Europeans revised their belief that swans are white. Psychology suggests, then, that reference is not stipulated in this way.

A belief's ordinariness, as I remarked in Section 3.2, is not something that can be discerned simply by looking at the "that"-clause we would normally use to attribute the belief. Something that in many ways appears to be an ordinary belief—"Gold is yellow" or "Something about swans causes whiteness"—might turn out to be, in virtue of a stipulation about reference elsewhere in the mind, irrefutable or extraordinary. This is to be expected: ordinariness is a matter of inferential potential, and inferential logic, like Newtonian gravity, connects everything to everything else. By the same token conceptual inductivism, in its attribution of ordinariness to a certain swath of beliefs, has implications not only for those beliefs but for the cognitive economy as a whole—implications that rule out the stipulativity of the simple theory of reference articulated above.

Move, then, to a more sophisticated theory of reference—say, the kind of theory inspired by Frank Ramsey, developed by Lewis (1970), and adopted by adherents of the "Canberra plan," in which a concept refers to whatever best satisfies a weighted mix of the "platitudes" in which the concept appears, and satisfies them well enough. The platitudes in question may be ordinary beliefs, yet in virtue of the way that reference is fixed, any term that succeeds in referring must satisfy some reasonably large set of those beliefs. The implications of a Ramsey-Lewis theory of reference for the vindication of philosophical analysis will be examined in Section 8.4; here my question is whether stipulating such a theory is consistent with conceptual inductivism.

On the one hand, stipulating a Ramsey-Lewis theory of reference allows for the refutability of any particular belief, since reference requires only the truth of a large set of beliefs, but not of any beliefs in particular. But on the other hand, it puts a constraint on the number of beliefs involving a given concept that may be concluded to be false simultaneously without the concept's being abandoned altogether, on the grounds of its failing to refer. Core causal beliefs and similar ordinary beliefs about a category's properties do not submit to any such strictures. It is logically quite possible to infer that everything you know about swans is false, while simultaneously seeking to learn the truth about that very category. Conceptual inductivism denies the existence, then, of the holistic constraints on belief that the stipulation of a Ramsey-Lewis theory of reference implies.

Is it possible that these constraints, which are not only rather esoteric but rather indirect, are simply ignored by the typical reasoner, constituting yet another case of mild but widespread irrationality in everyday thought? Or that the constraints are only very weak, and therefore barely noticeable in everyday cognition? I considered these ways of defending the existence of stipulative reference-fixers against conceptual inductivism in the discussion of explicit reference-fixing in Section 7.1; the responses I made there remain effective, I believe, even when the stipulativity is tacit or indirect. To recap:

the first response undermines the conceptual analyst's explanation of case certainty, while the second leaves very little to be discovered by analysis.

In any case, generalized conceptual inductivism suggests that the concept of reference is itself inductive. Regular thinkers may make many suppositions about reference—they may have a "folk theory" of reference—but there is nothing stipulative about them. They are ordinary beliefs that may be debated and weighed against alternatives just as in any other form of empirical or philosophical inquiry. Conclusions about membership of everyday, unsemantic categories might just as well motivate us to correct our theory of reference as vice versa (and indeed, in the philosophical analysis of reference, that is precisely what occurs). Our beliefs about reference, then, cannot serve as a source of stipulativity.

7.3 The Chalmers Defense of Analysis

The thesis that philosophical concepts are inductive challenges not only the viability of conceptual analysis, but also related views about the nature of armchair thought. As an example, let me show how it undermines a well-known argument developed by David Chalmers, which purports to establish that the kind of case judgments that drive philosophical analysis constitute a priori knowledge.[2] (And before I continue let me be clear that it is Chalmers's argument that is undermined, not his conclusion. I myself intend to argue in later chapters that philosophical case judgments are, even if not technically a priori, then at least reliably generated in the armchair. I simply don't think the goal we share can be reached by Chalmers's route.)

The ground for a case judgment, Chalmers proposes, can be understood as a material conditional whose antecedent is whatever properties are stipulated to characterize the case and whose consequent is the judgment itself. A judgment about a Gettier case, for example, would have the story about Sylvie, Bruno, and the copy of *Twilight of the Idols* (or whatever) as the antecedent and the judgment (canonically, that the belief in question is not knowledge) as the consequent. In philosophical analysis, Chalmers quite

---

[2] What I call the Chalmers argument in this section is the "argument from suspension of belief" in Chalmers (2012). Chalmers has a suite of arguments for conclusions of varying strengths; the one that follows, however, perhaps comes closest to the line of thought that he has advocated for many years—in Chalmers (1996), Chalmers and Jackson (2001), and elsewhere.

reasonably supposes, the aim of making case judgments is in fact to learn such "world-to-category" conditionals, which are then brought to bear on various hypotheses about the relevant category's essential nature, undermining some and supporting others. (I made use of this idea when discussing empirically informed analysis in Section 1.4.) Chalmers argues that in an important range of cases, the conditionals can be known a priori.

Take some concept, he urges; say, the concept of water. Put all empirical knowledge—that is, all knowledge justified even partly by way of experience—out of your mind, as though you know nothing of the way the material world is constituted. Abandon all belief, hypothesis, conjecture as well, about water or anything else. Now imagine a particular way the world might be. More specifically, imagine a particular set of fundamental-level physical facts.[3] If the world turned out to be that way, what would you count as water? Chalmers proposes that, under ideal circumstances, you will be able to answer this question for a wide range of specimens (perhaps all but a few borderline cases). Each of these decisions reflects knowledge of a "world-to-category" conditional with the imagined "way the world might be" as an antecedent and the water judgment—that the specimen is water or not water—as the consequent. Because no empirical beliefs play a role in the judgment—they are all, by fiat, suspended—your knowledge of the conditional is not based on experience, thus it is a priori. (For Chalmers, a proposition is known a priori just in case it is "known with justification independent of experience" [Chalmers 2012, 468].)

The story so far is not quite sufficient to vindicate philosophical analysis, since as Chalmers observes, a proposition that is known a priori on his definition might be empirically corrigible. It might, in other words, be known independent of experience but only defeasibly so, and therefore might be susceptible to empirical rebuttal or undermining. To justify the analyst's reliance on case judgments, then, it must be shown that the judgments are by and large undefeated (or at least that we are warranted in supposing so).

---

[3] Chalmers adds to the set three other kinds of facts: fundamental-level phenomenological facts, indexical facts that tell you where in space and time you yourself are, and a "that's all" fact to the effect that there are no further facts. The combination of the four he refers to as a *PQTI* basis. His argument is couched in terms of conditionals with some particular *PQTI* facts as antecedent and a fact about category membership as consequent.

7.3. THE CHALMERS DEFENSE OF ANALYSIS

The prime opposition here is the skeptical tradition in experimental philosophy; thinkers such as Machery (2017) might acknowledge the force of Chalmers's argument for aprioricity while maintaining that empirical research has shown that the dialectically most important case judgments are unreliable, thereby undercutting the a priori warrant for using them in the course of philosophical analysis.

In response, it might be argued either that the empirical evidence shows no such thing, or that the aprioricity of the conditionals is of the incorrigible variety. Chalmers holds out hope for the latter strategy. Suspend not only empirical beliefs, he urges, but all empirically corrigible beliefs (thus even corrigible a priori beliefs). Still you will find that you know many philosophically significant world-to-category conditionals. This knowledge must, then, be among our incorrigible knowledge.[4]

Serious challenges have been made to Chalmers's argument. Attempts to recover knowledge of the conditionals by "suspension of empirical belief," Block and Stalnaker (1999) argue, are typically defective because not all such beliefs are successfully suspended. We grasp the conditionals and make our case judgments, but we delude ourselves if we think that we have done so using only nonempirical, let alone incorrigible, beliefs.

Chalmers parries with a related argument—the "front-loading" argument. Choose some fact W about water; say, that the polar ice cap is made of water. Now choose some other facts C that, given what you know about the world, imply W. (C ought not to mention water explicitly, or the game is easy and uninteresting.) Then you should accept that you know the world-to-category conditional $C \to W$. Of course, there is liable to be much empirical knowledge founding your knowledge of the conditional. Put it all, Chalmers proposes, into the antecedent of the conditional. Even better, put all the fundamental facts S about the world into the conditional. Presumably you also know the new conditional $C \& S \to W$. Is there any further empirical knowledge E founding that conditional? If so, put it

---

[4] Chalmers argues for something stronger than incorrigibility, namely, what he calls "conclusiveness," which he takes to imply certainty (Chalmers 2012, 468). A belief might be incorrigible, in the sense that no *PQTI* fact could undermine it, yet uncertain—indeed wrong. This possibility will be an important element of my own view.

into the conditional too (this is "front-loading"). You surely know the revised conditional $C \& E \& S \to W$. No empirically derived facts ground your knowledge of the conditional, so it must constitute a priori knowledge. The process can go further: if there are corrigible a priori beliefs grounding the conditional, they too can be put into the antecedent, and *that* conditional will be known incorrigibly.

If conceptual inductivism is true, however, then none of Chalmers's arguments for our a priori knowledge of world-to-category conditionals can succeed.

Begin with the front-loading argument for the stronger conclusion: incorrigible knowledge of the conditionals. Consider again the conditional that underlies your judgment that the polar ice cap is made of water: $C \to W$. The facts in the antecedent C include, let's say, the proposition that the ice cap is made of H_2O. Your knowledge of the conditional hinges, then, on your knowledge that water is H_2O (or close enough). This knowledge is of course partly derived from empirical knowledge, such as your knowledge that the clear, potable stuff in the lakes and rivers around here is H_2O. Chalmers would have us put this and all other empirically founded knowledge in the antecedent of a new conditional $C \& S \to W$. On a traditional telling, knowledge of this conditional might be based on the knowledge that water is whatever has the same microstructure as the clear, potable stuff in the lakes and rivers around here. That proposition is the sort of thing that Chalmers expects to be known a priori and incorrigibly. He is not committed to the aprioricity of this piece of knowledge in particular; I use it as an illustrative example only. His conviction is rather that some piece of incorrigible a priori knowledge or other will turn out to ground the conditional built by front-loading all relevant empirical and other corrigible knowledge into the original conditional $C \to W$.

In the intensional analyst's view of the workings of the human mind, this posit would be vindicated. Explicitly or tacitly, the mind encodes some sort of reference-fixing rule for "water." That water is whatever has the same microstructure, etc., is precisely the sort of thing that might constitute such a rule, which would then because of its stipulative status qualify as incorrigible.

In the conceptual inductivist's view of the workings of the mind, however, Chalmers will have no such luck. The belief that water is whatever has

7.3. THE CHALMERS DEFENSE OF ANALYSIS

the same microstructure, etc., is an ordinary belief like any other, derived most likely from other ordinary beliefs such as the core causal beliefs about water and further posits (see Section 9.6 for a fuller story). Thus whether it qualifies as knowledge or not, it will be properly treated—given its content and its lack of any special semantic status—as empirically corrigible. More generally, every inferentially useful belief about water will be an ordinary belief, and so logically subject to empirical confirmation or refutation.[5] The only way to get from the antecedent of one of Chalmers's conditionals—a set of facts about the world that makes no specific mention of waterhood—to its consequent—a judgment of waterhood—is by way of beliefs about water. Thus the only way to get from antecedent to consequent is by way of ordinary beliefs, which are corrigible. Such conditionals are therefore known only corrigibly.

What if we were to front-load these beliefs, producing a new conditional with all of our water beliefs packed into the antecedent? The resulting conditional can be known a priori, I think, yet it contains no information about water, but rather expresses a general fact about inductive logic, namely, that from such and such empirical facts about a substance you might call "water" such and such a further fact about "water" would follow. For this very reason—the vapidity of the resulting conditional—Chalmers forbids the front-loading of propositions about water itself.

A similar line of thought shows that inductivism undermines Chalmers's belief suspension argument, with which this section began. Suspend your corrigible knowledge and beliefs about water, and what is left? If inductivism is correct, all your beliefs about water are ordinary beliefs and are therefore corrigible. Thus, once you are done suspending, nothing is left: you have no remaining beliefs, no theory whatsoever, about water. Consequently, you will be unable to make any case judgments, unable to come to know any world-to-category conditionals, about water. You will have nothing to go on. Your water concept, insofar as it exists at all, is an empty token "water" utterly inferentially isolated from the rest of your mind.

---

[5] An inferentially useful belief about water is one that might be used to categorize a specimen as water. Useless beliefs include tautologies such as "Water is either water or cheese" or other beliefs whose a priori justifications have nothing substantially to do with water, such as "Either three is a prime number or water is H_2O."

In making the case against Chalmers, I have so far relied on the corrigibility of ordinary beliefs. Consequently, I have not undermined the weaker versions of the arguments, which aim to establish the aprioricity of the world-to-category conditionals without making any claim about their incorrigibility. As I remarked above, the weaker versions are not on their own sufficient to vindicate philosophical analysis, but it is nevertheless salutary to sketch a way in which they might be blunted by conceptual inductivism.

The strategy I have in mind parallels the strategy used against the stronger versions above: in the case of water, to invoke inductivism to conclude that all inferentially useful beliefs about water are ordinary; to argue that ordinary beliefs about water are not a priori; and to conclude that suspending non–a priori beliefs leaves us without any useful beliefs about water, thus unable to know conditionals about water. The key premise, besides conceptual inductivism, is that ordinary beliefs are not a priori, which in Chalmers's terms means that they are not justified independently of experience. On the face of things, it certainly seems that ordinary beliefs about water must be empirically justified if they are justified at all. After all, they have empirical content: they assert that the world is configured in a certain way and they are refuted if it turns out not to be that way. That may be enough to clinch the argument against the aprioricity of water conditionals.

It is somewhat less clear that ordinary beliefs about philosophical categories such as knowledge are justified on the basis of experience. But to defuse Chalmers's argument, it is enough to raise doubts. If inductivism is true, then our knowledge of world-to-category conditionals is certainly based on ordinary beliefs, and it is far from clear that any ordinary belief is a priori. Thus, we should refrain from concluding that our knowledge of the conditionals is a priori.

7.4 Conceptual Inferentialism

My argument against Chalmers's vindication of philosophical analysis assumes that, even if we have a water concept, we cannot use it to make inferences about water without having some beliefs about water—beliefs that must be ordinary and therefore not incorrigible, if causal minimalism is correct. The inferentialist approach to semantics provides what is apparently a way around this assumption.

7.4. CONCEPTUAL INFERENTIALISM

Inferentialism proposes that many concepts by their very nature—in virtue of their essential semantic properties—warrant certain inferences without the assistance of generalizations, ordinary or otherwise. Consider the swan concept, for example. An inferentialist might suggest that anyone possessing the concept is by that very token warranted to move inferentially from the belief that Charles is a normal swan to the conclusion that Charles is white. No generalization is necessary to function as a major premise; the reasoner need not subscribe to the belief that all normal swans are white or that the swan essence causes feathers to whiten. The swan concept itself is sufficient to send thought winging on its way from swanhood to whiteness.

Such a reasoner might then put aside all their ordinary beliefs as Chalmers suggests, yet still make inferences about swans, and so perhaps judge certain specimens to be swans or not. More generally, such a reasoner is apparently in a position to gain a priori knowledge about swans: simply by possessing a concept, they accrue knowledge of conditionals—if this, then that—which when dressed up as case judgments, by way of philosophical analysis stand to illuminate the essential nature of the corresponding category.

This very consequence of inferentialism suggests, however, that it is at odds with conceptual inductivism for the same reason that tacitly stipulated intensions are at odds with inductivism. If it is inherent in the concept of "swan" itself that any normal swan may be reasonably inferred to be white, then the belief that swans are white cannot be, in those that have it, an ordinary belief. That is because it cannot be revised to reflect the fact that swans are sometimes black, on pain of incoherence: if it were, the reasoner would find themself in a position to infer justifiably from something's swanhood both that it is white and that it is maybe black. In short, if the concept of swan licenses inferences about swan color, then some beliefs about swan color are not ordinary beliefs. If causal minimalism is correct, inferentialism about the swan concept is false.[6]

Or at least, inferentialism about swan color is false. You might play around with more subtle inferential patterns, building their warrantedness into the swan concept to see what emerges. My bet is that inherent inferential warrant strong enough to be interesting will stand to confer

---

[6] What if the warrant for inferring whiteness is merely prima facie or pro tanto? Then the warrant consequently conferred on case judgments will be too weak to vindicate analysis.

extraordinariness on beliefs about swans that can be empirically demonstrated to be ordinary (a version of the "master argument" from Section 7.1). The mind might have been designed around an inferentialist semantics. But it is not.

7.5 Philosophical Analysis Is Not Conceptual Analysis

If armchair philosophy is largely pursued by way of modern conceptual analysis, then what we come to know through philosophical analysis are the consequences of certain (perhaps indirectly, perhaps tacitly) stipulative propositions about the category in question. Stipulativity implies incorrigibility of some sort. Inductivism concerning the philosophical concepts does not (unlike Quine's philosophy) deny the possibility of stipulative belief, but it hypothesizes that as a matter of human psychological fact, every substantial belief that we form concerning our categories, including the philosophical categories, is corrigible—not merely revisable (even definitions can be revised) but in a position to be supported or undercut by reasoning or evidence.

In short: modern analysis seeks something that conceptual inductivism says is not there. The same is true of intensional analysis and, for slightly different reasons, hypothetical analysis.

If conceptual inductivism is true of the philosophical concepts, then, neither modern nor intensional nor hypothetical analysis can yield any substantial facts about the philosophical categories—about their essential natures or about anything else. Philosophizing by reflection alone would be like checking the weather by reflection alone: you would draw a total blank.

But armchair philosophy does not draw a blank. Whatever the quality of our results to date, there is clearly plenty to engage philosophers without their having to look out the window. Something in the mind provides a basis for philosophical inquiry that is, if only prima facie reliable, then certainly extremely rich, generating metaphysical theories and putative essence postulates that are surprising, informative, and for the most part apparently accurate in their classifications—if perhaps always somehow flawed. These mental foundations, if they do not settle philosophical questions by uncovering stipulative, therefore conclusive, truths, must settle them . . . inconclusively? Inductively?

CHAPTER 8

Inductive Analysis

8.1 Inductive Analysis

In order for the classical analysis of a category to succeed, the corresponding concept must contain other concepts, the delineation of which contents will constitute the analysis. In order for the modern analysis of a category to succeed, the corresponding concept must be involved in one or more implicit definitions, or more generally representations that possess "stipulativity," the delineation of which definitions or stipulative beliefs will constitute the analysis. In order for the hypothetical analysis of a category to succeed, the corresponding concept must be built around an essence postulate, the delineation of which postulate will constitute the analysis.

If the preceding chapters are correct, our concepts of philosophical categories have none of this structure. They contain only ordinary beliefs—beliefs about the categories' characteristic properties, their causal or other explanatory roles, and any number of other things, but not beliefs that say something explicitly about the categories' essential natures and not beliefs that, because they are in a broad sense stipulative, say something indirectly or implicitly about essential natures.

Knowing all of this, we decide to push ahead with philosophical analysis as best we can. Selecting a philosophical category of interest, we formulate hypotheses about the essential nature of that category—hypotheses that a modern conceptual analyst would understand as attempts to capture the definition represented by the corresponding concept. We then test the hypotheses against our judgments about cases, looking for counterexamples (while attending also to the other traditional considerations, such as allowing intuitions to be explained away under certain circumstances and giving some weight to simplicity and unity).

If we believe that our concepts are inductive, and that our beliefs about the corresponding categories are merely ordinary, can we do any of this in good conscience? What weight can we give to our "intuitions" or judgments about cases—our judgments, that is, about membership of philosophical categories—if we do not have some definitive criterion for category membership ensconced in our heads? How, in particular, can we take so seriously our judgments about the complex, fanciful edge cases that figure in much philosophical analysis?

Here is what we might say to ourselves. Although our philosophical concepts are not susceptible of conceptual analysis, the ordinary beliefs in which they figure provide powerful heuristics for judging category membership, just as a minimalist basic natural kind concept provides, in the form of its associated core causal beliefs, powerful heuristics for judging membership of the kind. Our classifications under the philosophical categories are therefore accurate: we know singular causation, justice, and so on when we see them. Consequently, the analysis that we concoct and refine in the course of philosophizing about a category can be expected to come closer and closer to capturing the category's extension, and with any luck, closer and closer to capturing the true criterion for category membership, that is, the category's essential nature. Such analyses do not assert facts about the mind; they are not in any substantial sense "conceptual." But they do assert increasingly well-tested hypotheses about essential natures. As philosophical analyses, they may be a complete success.

I will call philosophical analysis that works in this way *inductive analysis*. It has two phases, each largely driven, as the name proclaims, by inductive reasoning.

In the first, descending phase, our ordinary beliefs about a philosophical category are used to make decisions about individual cases: Is this a cause? Is that knowledge? As with our natural kind classifications, these decisions are not based on posits or stipulations about a category's essential nature, but on nonessential connections between the category and the world that

8.1. INDUCTIVE ANALYSIS

are rich and strong enough to imply which things fall into the category and which do not. Just as a West Coast tree-spotter identifies California nutmegs using core causal beliefs that do not articulate the essential nature of the species, so the beliefs we use to classify (say) a Gettier case as nonknowledge are beliefs about knowledge but not beliefs about the nature of knowledge. Like core causal beliefs about natural kinds, they may articulate explanatory connections—say, between knowledge and other epistemic properties or between knowledge and truth (Sections 6.5 and 13.4)—but they do not take the further step of attributing to these deep relations defining power. Regardless, they make the distinctions we need.

In the second, ascending phase of inductive analysis, case judgments about category membership are brought to bear on hypotheses about the category's essential nature. The judgments lend support to those hypotheses that predict them and count against those hypotheses that predict their absence. Considerations of simplicity, unification, and plausibility may enter into the analyst's deliberations in the ascending phase, as they do in scientific reasoning or other forms of ampliative inquiry, such as the interpretation of sacred texts or the reading of people's moods from their facial expressions. They are not, however, strictly necessary: the theory that best predicts the judgments about cases may be declared the winner on those grounds alone, and so may be counted as our best stab at the philosophical truth.

With respect to the descending phase, inductive analysis and modern conceptual analysis agree that case judgments or "intuitions" are regulated by the corresponding concepts: it is what is in your knowledge concept that moves you to count one thing as knowledge and another thing as not. They disagree on the logical status of the regulator: for conceptual analysts, the concept stipulates the ultimate ground of category membership—the essential nature of the category—whereas for inductive analysts, the concept contains only ordinary beliefs about the category, profound and far-reaching though their implications may be.

Whereas the difference in the descending phase is not one of where you get to but of how you get there, the difference in the ascending phase is the reverse. In inductive analysis, case judgments are evidence for and against philosophical theories, that is, theories about the essential nature of the category in question, that bear upon these theories in ways characteristic of the evidence-based reasoning found in science. In modern conceptual analysis, the reasoning—at least when the analyst is self-conscious—works in much the same way, but the hypotheses to be tested are not philosophical but psychological: they are hypotheses about what is in the concept, and thus hypotheses about the structure of the representations that generated the case judgments. The ascending phase of inductive analysis looks scientific, but the corresponding phase of self-conscious modern conceptual analysis *is* scientific: it has the same aims and methods, proximally, as cognitive psychology.

The ultimate aims of modern conceptual analysis are, however, philosophical rather than psychological. So there is in modern analysis, practiced self-consciously, a third phase following the ascending phase in which philosophical conclusions are inferred from psychological premises—in which, to be more exact, the essential nature of a philosophical category is taken to be exactly what is stipulated by the corresponding concept. In this way the modern analyst completes the journey from head to world that, for the inductive analyst, was finished at the end of the second phase.

In the course of modern conceptual analysis, then, the relevant philosophical concept features twice: first as the generator of the case judgments, then again as the subject matter of the hypotheses on which those case judgments are brought to bear. Self-conscious modern analysts not only think with their concepts, they then proceed to think about them.

In the course of inductive analysis, the concepts do not enter into a self-conscious analyst's deliberations at all—any more than in the course of normal research a physicist deliberates about their electron concept or a biologist about their concept of DNA. The inductive analyst uses their judgments about category membership to make judgments directly about the nature of the category itself.

Inductive analysis has much in common, then, with the account of philosophical methodology urged by Williamson (2007). Like the inductive analyst, Williamson rejects the view that philosophy takes conceptual structure as its subject matter; he goes on to argue that philosophical thinking is an application or modest extension of our "ordinary ways of thinking" to philosophy's real subject matter, the nature of the philosophical categories. This view is entirely consonant with the view that philosophical analysis is inductive analysis, and that Williamson and I both use the word "ordinary" is, I think, no coincidence.

8.1. INDUCTIVE ANALYSIS

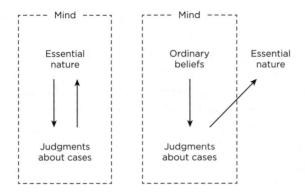

FIGURE 8.1. Conceptual analysis (left) versus inductive analysis (right).

A great advantage of the inductivist picture, above and beyond the modern picture's incompatibility with our most advanced psychology of concepts, is that, while modern conceptual analysis guides the analyst around an inferential circuit leading back into their own head, inductive analysis offers the prospect of genuinely new knowledge concerning the world outside. The contrast is illustrated pictorially in Figure 8.1. Descending arrows in the diagram represent the process of theory-driven categorization, that is, the use of a theory to determine category membership, and ascending arrows represent the inference from judgments about particular cases to philosophical theses about categories.

In modern conceptual analysis, the starting point—the ground for making case judgments—is a mental representation of a category's essential nature. The end point is the very same assertion about essential nature, now understood as a philosophical thesis about the category itself. The knowledge about the philosophical categories that is delivered by modern conceptual analysis is, then, nothing more than an expression of some private stipulation that you have, apparently unbeknownst to yourself, made inside your skull's cave, rather than a reflection of the configuration of things out there beyond the skin.

The inductive analyst paints a sunnier picture.[1] Their starting point is a set of ordinary beliefs that are rich and various enough to provide a reliable

---

[1] This picture will be considerably complicated by developments in Chapters 11 through 15. But I will ultimately conclude that the virtue of analysis is its ability to give us knowledge of the explanatory structure of the world outside.

basis for categorization in a wide range of cases, but that do not pretend to represent the underlying nature of the category itself. The end point is a hypothesis about the category's essential nature that is not antecedently encoded in the mind and that consequently does not serve as the ground for judgments about cases. It is something wholly new; it is (if correct) fresh knowledge about the ultimate basis of category membership.[2] Philosophical analysis thus constitutes, in the inductivist's view, an investigation of a question whose answer is not already spelled out in some dimly lit cognitive fissure, but rather must be inferred ampliatively from what is previously known. The results of analysis are surprising, then, because they come from outside the armchair into the mind, and they tell you something about the world that you did not, in the course of constructing your personal mental dictionary, dictate yourself.

8.2 Inductive Case Judgments

The remainder of this chapter discusses a series of objections to the thesis that philosophical analysis is inductive analysis.

In the inductive picture, case judgments are characteristically made, as in the case of basic natural kinds, by way of inductive reasoning using ordinary beliefs. There are two reasons, I think, why you might resist the inductivist understanding of philosophical case judgments.

The first is adherence to what I called in Section 6.4 the "myth of depth": the assumption that deep thought, such as the sophisticated deployment of philosophical categories, requires metaphysically deep representation, that is, representation of categories' essential natures and other foundational truths.

Chapters 5 and 6 aimed to shatter the myth. Successful and subtle thought about basic natural kinds, such as folk genera and chemical substances, can proceed without any metaphysical commitments; all that is needed are, I showed, explanatory commitments linking kind membership to observable properties. Our philosophical case judgments can, I propose, be as reliable, and so can give us as sure grounds for philosophical analysis,

[2] It is in principle possible for inductive analysis of a category to show that the ordinary beliefs grounding the case judgments also capture the category's essential nature, although learning this would all the same count as new information about the category.

8.2. INDUCTIVE CASE JUDGMENTS

as our judgments about basic natural kind membership—while being just as "inductive."

The second reason to suppose that armchair case judgments cannot be inductive is more searching. I introduced, in Section 1.3, the notion of case certainty: the thought or feeling that our judgments in some philosophical thought experiments are incontrovertible. That the justified true belief in the Gettier scenario is not knowledge; that the presence of backups cannot rob events of their status as causes—these conclusions seem decisive, apodictic, indisputable.

Inductive thinking, however, is provisional. Its conclusions stand to be overturned by new information, including new empirical information. When we classify something as a swan, for example, we know that we might learn more about the specimen, or more about swanhood, that would cause us to retract our judgment. Our concept of swanhood is "only a theory"; judgments made on the basis of the ordinary beliefs that constitute the concept's cognitive structure are therefore eminently corrigible. How, then, to explain case certainty?

Even the modern conceptual analyst, though they seemed well placed to shape a convincing explanation, had difficulty in draping the Rawlsian veil so as to obscure definitions while allowing their authority to shine through. As a consequence, I considered the possibility that case certainty is a philosophical methodologist's conceit—that ordinary working analysts do not regard even their clearest case judgments as certain, but rather as "settled enough," in the same way that we regard perceptual judgments as "settled enough" in situations where, regarding their veridicality, there is no prima facie cause for concern (Sections 3.2 and 3.3). An inductive analyst might make the same move. But I have a higher ambition: I aim to explain how even ordinary working analysts with inductive concepts might come to regard certain case judgments as incorrigible.

Case certainty, as I remarked in Chapter 1, might come in either of two ways. First, I might believe that I have the evidence I need to make a conclusive judgment. Further evidence might exist, but it could not possibly overthrow my conclusion. Second, I might not feel that the internal logic of my judgment itself invests the conclusion with certainty, but I might nevertheless be confident in the judgment while also believing that I have in my possession all the evidence relevant to making it. In both situations, I view my judgment as unshakable: no further information could undermine or rebut it. But in the second situation, I reach this judgment by reflecting on my epistemic position, and not only on the logic of my reasoning.

The first, "strong" way is largely closed off to the inductivist: the ordinary beliefs are typically not certain, and even if they were, that certainty would not be transmitted by the inductive reasoning through which many categorizations, especially positive categorizations, are made. But even an inductivist can reach case certainty in the second, "weak" way, and that will be the foundation for my inductivist explanation of case certainty in Chapter 14. There will be a few sharp curves in the road, however, so I will promise nothing more at this early stage.

8.3 Armchair Discovery

Knowledge of the world outside the head can be gained by armchair rumination, if the inductive view of philosophical analysis is correct. Is that not impossible?

Think of one of the great detectives of fiction—Holmes, or Drew, or Lönnrot, or Wallas—settling down at the end of the day and contemplating all the evidence they have gathered over the course of the previous chapters. Nowhere in that evidence is the identity of the murderer, yet by the time the detective is done thinking, they know who committed the crime. Far from being miraculous, the ability to learn new things in the armchair is the characteristic attribute of inductive thinking, the feature from which ampliative inference takes its name. What is discovered in this way is not, of course, conjured from nothing: the detective takes a substantial knowledge of the relevant empirical facts with them to the living room. From empirical knowledge the human mind spins more empirical knowledge.

If philosophical analysis is inductive analysis, then the mechanism by which it creates knowledge in the ascending phase is nothing special or proprietary to philosophy. The case judgments are evidence, and they function in philosophical reasoning by way of relations of inductive support or undermining.[3]

---

[3] Sometimes, as in science, the undermining can be deductive rather than inductive; just as the white raven refutes *All ravens are black*, so the Gettier cases refute the justified true belief theory of knowledge. The ascending phase of inductive analysis is not purely inductive, then, any more

There are, however, two ways in which inductive analysis differs from forensic reasoning, or more generally from scientific theorizing, that might deter you from accepting that novel philosophical facts can be discovered in the armchair.

The first is the mysterious nature of those facts themselves. What is the source, the ground, of truths about a category's essential nature? If conceptual analysis in all its forms is rejected, it cannot be a definition or other semantic fact stipulated, explicitly or tacitly, by the inquirer. What, then, props up reality—what supplies the standards by which a thesis about some essential nature qualifies as right or wrong—in that awkward period during which the mind has yet to settle on a final answer?

For now, I ask you simply to suspend disbelief. Given case judgments and inductive reasoning, we can conduct an orderly and systematic test of hypotheses about essential natures. Postpone, without forgetting, the question of the ultimate subject matter of these hypotheses. The topic is to be taken up in Chapters 9, 11, and 15. Some essential natures, I will conclude, are features of external reality determined by the facts about reference; some are in a significant sense mental constructions; some will turn out not to exist at all.

The second disconcerting difference between inductive analysis and empirical reasoning lies in the premises. In scientific inquiry the world, properly cajoled, yields the facts that propel us inductively to deep truths about nature. In murder mysteries the witnesses chime in. But in inductive analysis our own minds, working alone, provide the case judgments that play this evidential role. Can case judgments, so detached from the world out there, be trusted as sources of evidence about its content and structure? The next section continues my tale of our inductive voyage across the armchair/world barrier by tackling this question of reliability.

8.4 The Importance of Having a Good Start in Life

The epistemic status of a conclusion reached by a successful session of philosophical analysis depends on the initial commitments that send it on

---

than scientific reasoning is purely inductive. Still, the classification of both as species of inductive inquiry, broadly speaking, seems apt enough.

its way, namely, the ordinary beliefs about a philosophical category that guide the case judgments that "confirm" or "disconfirm" our hypotheses about the category's essential nature. It would seem to be necessary, if the process is going to produce justified beliefs, that the ordinary beliefs are reliable or justified or have some other positive epistemic valence.

Suppose I start out at the very beginning of my philosophical career, for example, with some ordinary beliefs about causality. These are the beliefs that guide my judgments about the various problem cases—late preemption, double prevention, and so on—that figure so prominently in the recent philosophical dialectic about singular causation, as well as guiding, less prominently but no less importantly, those judgments about unproblematic cases that a theory of causality must replicate to qualify even as a prima facie candidate for serious philosophical discussion.

Such a discussion is possible because I find that my fellow philosophers more or less concur with my judgments of causality, citing grounds for these judgments similar to my own. Take a case of late preemption: Bruno and Sylvie are throwing furniture at a store window; Sylvie throws first and her stool shatters the window, while Bruno's chair, also on target, arrives a few moments later and encounters just the empty space where the window used to be. We agree that Sylvie's throw was a cause of the window's destruction and that Bruno's was not. Further, we agree that the fact that Sylvie's stool physically contacted the window is crucial to its being a cause, and the fact that Bruno's chair did not contact the window is crucial to its not being a cause. (We do not, however, necessarily regard the grounds that we cite—in this case, the matter of furniture/glass connection—as directly revelatory of causation's essential nature. Many of us would indeed deny that contact is either necessary or sufficient for singular causation in general.)

We share some general beliefs about causation, then, and apparently, we apply those beliefs in much the same way, coming to much the same judgments about particular cases. And yet a still, small, skeptical voice whispers in our philosopher's inner ear: how do we know that these beliefs and these judgments are at all justified, reliable, truth-directed, or in fact bear any relation whatsoever to causal reality?

Of course we treat them that way, confidently making judgments of causality for everyday, scientific, and legal purposes with a clear epistemic conscience. You might say, in a Williamsonian spirit, that if these judgments—many of which have great practical, theoretical, and moral importance—are good enough to trust in real day-to-day life, then they are surely good enough for philosophy. That is absolutely correct, I think, but not very philosophical, or at least not philosophical enough for a book on philosophical methodology. I want to know *why* the judgments are trustworthy. I would not be taking my job seriously if I were not a little worried.

What are the options for vindicating the "starter beliefs"? The first is what might be called the *empirical strategy*, an attempt to amplify in argument form the previous paragraph's *sangfroid*. We have been successful enough so far, goes the story, in applying the ordinary beliefs about causation and the rest that we found in our heads when we set out on life's journey; those beliefs must, then, be getting something important right. If you suppose that they are put in place by natural selection, the argument becomes all the more powerful for the beliefs' tempering in rivers of Darwinian blood.

Naturally, the empirical strategy with its evolutionary flavor works best for those philosophical categories that have immediate practical applications in hand-to-hand combat or late-night maneuvers in the boudoir—the kind of stuff on which human survival and proliferation most directly and obviously hinge. These days we can point also to the technological benefits of science, as realized in our flying machines and personal communicators, which surely speaks almost as strongly as our genetic momentum in favor of our judgments about causality, space, time, species, substance, and mental states. The prospects for the empirical strategy with respect to the good and beautiful are less clear (Joyce 2006; Street 2006). But let me push ahead with the more material categories in mind.

Against the empirical strategy its critics have deployed a simple, brutal fact: a system of beliefs need not be truth-directed to be useful.[4] Consider, for example, our folk or naive theory of physics, that is, the set (or sets) of rules we use, untutored, to predict the physical behaviors of things in the world around us. We are fairly proficient with physical stuff, in great part because it tends to do roughly what we expect it to do. There is much of practical importance, then, that naive physics gets right. But it is quite plausible that the tenets of naive physics, insofar as they have theoretical

---

[4] For other objections to the empirical strategy, see Stich (1990), chap. 3.

content, are largely mistaken. Even without undertaking the appropriate psychological investigations, you can guess that naive physics knows nothing of quantum or relativistic effects. Worse, there is good evidence that it is not even Newtonian: educational researchers asking why high-school and introductory college physics is so difficult to learn have suggested that naive physics has something of the character of medieval impetus theory (Clement 1983; McCloskey 1983; Kozhevnikov and Hegarty 2001).[5] This seems far from impossible: it might be that a theoretically deeply off-kilter theory makes predictions that are in the relevant contexts quite good enough for utilitarian purposes. Such a theory has great survival value, but limited epistemic value.

Is that reason enough to doubt the empirical strategy for justifying inductive analysis? It is fairly good reason to doubt the strategy as a justification of hypothetical analysis, which reads the essential nature of a category more or less straight off a belief about that nature.[6] But inductive analysis invokes our theoretical beliefs in the first instance only as categorizing principles. What matters for the epistemic standing of inductive analysis is that the case judgments it brings to bear on hypotheses about essential natures are reliable or justified in the right sort of way. Thus what matters is that the ordinary beliefs that guide those categorizations make good categorizations, not whether they are true. The present objection to the empirical strategy does give us some reason to doubt the beliefs' truth, but as developed so far it gives us no reason to doubt the accuracy of the categorizations based on the beliefs—just as in the case of physics, we have reason to doubt the principles of naive physics but no reason to doubt the accuracy (within practically relevant margins of error) of their predictions.

There is a disanalogy with the case of naive physics, however, that revivifies the objection. The predictions of naive physics about objects' behavior are tested directly against the world: typically we get to see whether or not things move in accordance with the theory's forecasts. Categorizations (and

---

[5] It is not clear that the impetus-centered intuitions elicited by these studies are the same principles that snatch us from the sabertooth's jaws: they might be purely post hoc attempts to explain predictions made by some consciously inaccessible part of the mind. In that case, what follows is more a parable than a true story.

[6] For an illustration of the "more or less," see the sketch of the Weathersonian view in Section 3.3.

this is true both of philosophical categories and of the basic natural kinds such as swans and water) are not directly tested in this way. We cannot measure the accuracy of our theory of swans by comparing its judgments about swanhood against some answer key—some master list of category membership—that nature provides for our convenience (Cummins 1998); the same is true of our judgments about what is and is not a cause. These case judgments play their part in our dealing successfully with a causal, swan-ridden world rather by facilitating inferences that lead to appropriate behaviors in such a world, such as avoiding aggressive waterbirds.

But just as false beliefs (of the right kind) about physical principles can lead to correct predictions about physical behavior, so incorrect categorizations (of the right kind) can lead to appropriate behaviors. We might classify tigers as swans and swans as tigers, but as long as the same actions are appropriate with respect to both kinds of animals, this wholesale error will leave us no worse off than the most exemplary taxonomy.[7] When the error concerns just a few borderline cases, things look even worse. It is far from clear that, say, going one way rather than the other on the question whether Gettier cases constitute knowledge or whether double prevention constitutes singular causation will make much of a difference to our prosperity or proliferation (especially since we have alternative ways of thinking about these cases in terms of justification, prevention, and so on). Yet such judgments figure crucially in our search for the essential nature of knowledge and causality.

The empirical strategy for vindicating our judgments of philosophical category membership, like the empirical strategy for vindicating our ordinary beliefs about the philosophical categories, therefore falls short. It is not terrible, but it is also not terribly convincing.

How else might we secure some connection between our beliefs about philosophical categories and the truth? There are the two hallowed philosophical strategies for understanding armchair knowledge, introduced in Chapter 1 and put to work to achieve consonance of category and concept in Chapters 2 and 3: nativism and constructivism.

On the nativist approach, some great and benevolent power has gifted us with beliefs that accurately reflect the nature of things out there in the

---

[7] Compare the argument against the empirical strategy in Williamson (2007, §8.2).

real world. What might that power be? It might be evolution by natural selection. If so, the objections lodged above against the empirical strategy apply: we can reasonably expect natural selection to equip us for success, but it does not follow that it has equipped us also with the truth, at least with respect to matters of fact that are not exposed directly to nature's teeth and claws but function rather as inferential middle terms.

And that other great benevolent power? He has his own problems (Nietzsche 1887, §125).

That leaves the constructivist strategy, which is in its loosest form the idea that the structure or mental content of a concept plays so central a role in determining the nature of the category to which the concept refers that thought concerning the category has some inherent tendency to truth.

In philosophy based on the classical theory of concepts, the conceptual structure in question takes the form of a mental definition, which wholly determines the essential nature of the corresponding category. The tendency to truth is strong and simple: the propositions that make up the definition are guaranteed to be correct. Any reasoning based on those propositions, then, has as sound a foundation as could possibly be desired.

A weaker philosophical view of the same orientation is implicit in Lewis's (1970) Ramsey-style theory of reference, sketched in Section 7.2, according to which the reference of a concept is determined by the platitudes in which the concept centrally figures: the category picked out by a concept is, in this view, the one that best satisfies the platitudes, provided that it attains some minimal standard of satisfaction (or else the concept fails to refer). It follows that, although no particular platitude is guaranteed to be correct, for non-empty concepts some minimum number of platitudes must hold true. If you suppose, for example, that for reference to succeed, "most" of the platitudes must be satisfied, then you can be sure that if the concept refers at all, most of what you platitudinously believe about the corresponding category is more or less on the mark.

I doubt that many philosophers would classify Lewis as a constructivist, and so I will allow that you can take advantage of the constructivist explanation of armchair knowledge without going quite so far as to endorse anything deserving of the name constructivism. What matters is that some of our beliefs about a category help to determine which category the beliefs are about. Call this the *reflexivity* of reference or belief. The constructivist strategy is to explain the reliability of certain armchair beliefs by invoking their presumed reflexivity.

This is precisely what I propose to do. To implement the reflexive strategy I must, as an inductivist, find a theory of reference in which the ordinary beliefs that make up a philosophical concept play a part in determining its reference, and show that the beliefs can for this reason be expected to be, on the whole, reliable—or something like that.

Two difficulties present themselves. The first is a special problem for the inductivist: just a little reflexivity can make the ordinary beliefs extraordinary, and so contradict inductivism itself. My earlier discussion of the Ramsey-Lewis approach to reference in Section 7.2 provides a salutary example, showing that even the nonspecific, holistic reflexivity implied by that theory transforms certain consequences of apparently ordinary beliefs into semantic necessities, which is incompatible with the inductivist way. What is wanted is a variety of reflexivity that allows for the falsehood of any or all ordinary beliefs, while nevertheless establishing those same beliefs as a reliable starting point for philosophical analysis.

The second problem is perhaps more perplexing still. There is no free lunch. *Ex nihilo, nihil fit*. Certainty is inseparable from nothingness (to repeat Reichenbach's aphorism from the first chapter). In other words, the reflexologist gets reliability, but at the same time renders philosophical analysis unable to yield anything genuinely new and interesting.

In Section 2.3 I rehearsed the worry that modern conceptual analysis must fail to provide substantive philosophical knowledge. Stipulate that a swan is a white bird and you are guaranteed that all swans are white, but you do not thereby learn anything new about the world. Out there the birds go about their business, chromatically unconstrained by your definition-mongering. Indeed, you do not get anything out of the process of analyzing your definition that you did not yourself put in by hand. Definition-driven reflexivity can supply novel conclusions, then, only if you have selectively forgotten, thanks to some Rawlsian anti-mnemonic aid, that you stipulated things to be that way in the first place, or because you have failed to grasp some deductive consequence of your stipulation (Balcerak Jackson and Balcerak Jackson 2012).

It is not true of inductive analysis that it extracts only what has already been written in the head. As I have explained above, novel knowledge of essential natures is attainable in the armchair through ampliative reasoning. All the same it is reasonable to worry that if such knowledge is given a reflexive foundation, it may turn out to be, however new, not truly substantive.

Here is an agenda, then: find a kind of reflexivity that is consistent with inductivism—the goal of the next chapter—and then show, after a near-death experience in Chapters 10 and 11, that it empowers inductive analysis to plumb matters of substance—the goal of Chapters 12, 13, and 14.

CHAPTER 9

Reference

9.1 Reflexivity versus Objectivity

If your beliefs play a sufficiently self-regarding role in determining their own subject matter, they must be somewhat directed toward the truth. Acquire a concept that refers in this reflexive way, then, and by a sleight of semantics you secure a lead in acquiring knowledge of the corresponding category without the least epistemic due diligence. To vindicate inductive analysis, I will appeal to the reflexivity of reference to establish that our ordinary beliefs about philosophical categories are trustworthy starting points for the endeavor.

It is not hard to find reflexivity in modern theories of reference. As I noted in the previous chapter, in a Ramsey-Lewis theory, according to which a concept refers to whatever category satisfies a good proportion of the platitudes involving that concept (provided that it satisfies more of them than does any other category), you can be assured that a good proportion of such beliefs—though you may not know which—are true.

Further, any theory of reference is partly reflexive if it incorporates a principle of charity, that is, a principle that puts some weight, when determining reference, on the number of beliefs made true by a choice of extension. (In Williamson's [2007] variant, the knowledge maximization principle, what is valued is the number of beliefs that are rendered not only true but known to be true.)

Reflexive theories of reference may be convenient and reassuring, but are they plausible? Casual reflection suggests, and the thesis of conceptual inductivism apparently insists, that it is possible for a thinker to be almost completely mistaken about the nature of natural kinds and other such "worldly" categories. Isn't the possibility of large-scale error incompatible with a reflexive approach to reference? To be wrong about a category we must be able to think about it—our concept must refer to the category. Reflexivity, however, seems to rule out a scenario in which we are abundantly wrong about a class of things to which we can successfully refer. Thus, the possibility of error implies that whatever the theoretical advantages of reflexivity, it is not by a reflexive relation of reference that our concepts, or many of them, connect to the world. Without further argument, then, we cannot fall back on reflexivity as a reason to trust our philosophical beliefs.

I might continue in this vein, treating the possibility of massive error as an ugly fact standing in the way of otherwise appealing reflexive conceptions of reference. But there is something deeper to say. The possibility of error is not merely real, it is supremely desirable. We should seek out concepts that allow our beliefs to blunder into rank falsehood, and we should be contemptuous of theories of reference that promise to lay too easy a road to truth.

The reason, in a word: substance. We want knowledge of essential natures to be substantial rather than trivial or empty, and as I argued very loosely in Chapter 1, there is reason to suppose that substantiality is attainable only if our concepts are "open" rather than "closed." A open essential nature oozes into the world like liquid wax and takes on whatever shape it finds there; its final contours must therefore be discovered through concerted inquiry, and so constitute a subject matter concerning which we could be, in the early stages of our investigation, quite mistaken.

Or to put it another way, a substantial essential nature implies an objective category, that is, a category whose boundaries somehow transcend the naive beliefs that are our only tools at the beginning of inquiry. If it is for logical or semantic reasons impossible for my initial beliefs to be mistaken about a category, then I must have constructed for myself a concept that imposes by fiat a certain scheme of classification on the world, that is, a

9.1. REFLEXIVITY VERSUS OBJECTIVITY

closed concept. That scheme might by chance correspond to an objectively correct scheme, but most likely it does not—and if not, it never will.

Closed concepts do not cut me off from knowledge of objective categories. I can always reason my way, using some other concepts, to the second-order conclusion that my original concepts are of categories that I have imposed on, not learned from, the world. But analyzing the original concepts—determining their essential natures—would no longer seem an appealing avenue of inquiry. Rather, I ought to regard those concepts as primitive representational tools to be discarded or consigned to the museum—mental hand axes, where laser cutters and scalpels are wanted.

In short: we ought to treasure corrigibility in our beliefs because we ought to treasure conceptual openness. Let us welcome our susceptibility to massive error about the basic natural kinds, then, and let us hope for as much fallibility as possible concerning philosophical categories such as causality and knowledge. That hope must be mixed with foreboding, however, as long as massive error seems to be at odds with referential reflexivity, my preferred strategy for securing some warrant for relying on our "starter beliefs". My next business is to tighten the focus of the question whether reflexivity and massive error are compatible, by posing it for the case of the basic natural kind concepts.

The possibility of massive error concerning the basic natural kinds suggests that the reference of our natural kind concepts and terms cannot be more than minimally reflexive. The Ramsey-Lewis theory of reference, for example, is in its most straightforward form too reflexive to get the facts about natural kind term reference right: it implies that our core causal beliefs about kinds must be mostly roughly true, but it seems that they can be totally off the mark.

Consider, as an illustration, the following case, an elaboration of a scenario that I used to problematize the prototype theory of concepts in Section 5.3. Wandering one foggy day in the fens, you come across a new kind of waterbird. These somewhat swan-shaped creatures are pink, have blue bills, a tromboning call, and are remarkably inept at aquatic maneuvers. You reason that they are members of a species until now unknown to you; you decide to call them "schwanns." I take it that, at roughly the same time as this new word is implanted in the language, a new concept sprouts in the mind—the schwann concept. You go about your day, thinking, talking, marveling about the schwanns.

If causal minimalism is correct (what follows does not turn on minimalism, but it is convenient to work with a specific theory of basic natural kind concepts), then your schwann concept gets its cognitive significance from a set of core causal beliefs: something about schwanns causes pinkness of the feathers; something about schwanns causes blueness of the bill; and so on. According to the Ramsey-Lewis theory, the concept will refer to whatever class of birds best satisfies these beliefs, provided that it satisfies sufficiently many of them—provided, let's say, that it satisfies at least half of them.

It is conceivable, however, that more than half of your initial core causal beliefs about schwanns turn out to be false. Perhaps the group of schwanns you first encountered are affected by some powerful environmental contaminant that turns the local schwanns into blue and pink tromboners, when in fact normal schwanns are green with yellow bills and have a call that is more redolent of a French horn. It seems that we could, and indeed that in such cases we would, coherently think such thoughts as "I was wrong in almost all of my beliefs about schwanns." The Ramsey-Lewis theory does not allow this belief to be correct.[1]

The same goes for cases of sexual dimorphism, such as the microscopic parasitical male barnacles mentioned in Section 5.2, or dramatic reinterpretations of the fossil record. The Cambrian organism *Anomalocaris* was,

---

[1] Some ad hockery can save the Ramsey-Lewis theory. In any sophisticated version of the theory, some beliefs or platitudes will count more heavily in reference determination than others. A Ramsey-Lewis proponent might hypothesize that the core causal beliefs are weighted very lightly, and other beliefs—beliefs about, say, the category membership of the specimens with respect to which the term was introduced—are weighted very heavily, in effect tweaking the parameters to get something from Ramsey-Lewis that is as close as possible to the causal-historical account of reference. (For some more sophisticated philosophical engineering in this vein, see Jackson [1998], 37–41] and Weatherson [2003].) My aim here is not to criticize the Ramsey-Lewis theory but to make a broader point about reflexivity and error; thus, I will not consider these defenses and I will not claim that the theory is refuted.

for example, first thought to be a kind of shrimp; it is now understood to be a huge, leaf-shaped trilobite-ravener (Gould 1989).[2]

It is for these kinds of reasons that the causal-historical theory is preferred to the Ramsey-Lewis and other descriptivist theories as an account of the reference of basic natural kind terms (Kripke 1980; Devitt 1981). In one version of the causal-historical story, reference is fixed by the intention of the coiner that a term refer to a category exemplified by some baptismal specimen. In the case of the schwanns, then, the reference of "schwann" is fixed by my intention that the new term should refer to the category exemplified by the pink-feathered, blue-billed tromboners—whether or not my beliefs about the category at that or any later time are correct. Further, provided that my neologistic aspirations are realized, the reference is retained as my word is passed from person to person; whatever the beliefs of these new users of the term, it continues to refer in accordance with my original referential intention.

The great advantage of the causal-historical account, for the seeker of objective categories, is that it permits us to make profound and sweeping errors about the objects of our thought—and in so doing, it frees our categories to home in on objective worldly structure without running afoul of self-imposed semantic constraints. Both the good and the bad are possible because the causal-historical account is far less reflexive than descriptivist theories such as Lewis's riff on Ramsey. Indeed, it has seemed to many philosophers that the best theory of reference for basic natural kind concepts is the least reflexive: it is the theory that when determining the reference of a term or concept puts as little weight as possible—perhaps none at all?—on our beliefs, now or later, about the corresponding category. If the same were true for our philosophical concepts, there would seem to be little hope of using reflexivity to justify the analyst's relying on their philosophical starter beliefs.

Is there a golden mean to be found between the Ramsey-Lewis account, conferring reflexivity and so starting you out in life with truth for free, and the causal-historical account, setting you free to find objective truth? Perhaps what's wanted is a theory of reference that endorses a principle of

---

[2] Our understanding of *Anomalocaris* continues to evolve; it may be a worm-sucker that posed no danger to trilobites whatsoever.

charity—assign extensions so as to maximize truth or knowledge—but that allows charity to be trumped on occasion by other concerns? Williamson's knowledge maximization principle is claimed, for example, to secure a modicum of confidence in the accuracy of our ordinary philosophical beliefs while allowing for extensive error.

Such ideas offer only a tepid warranty for our beliefs about philosophical matters: if you are the kind of person who is inclined to think that such beliefs are among the shakier of our doxastic commitments, then even as you sink into your professional upholstery you will be nagged by the thought that a global principle of knowledge maximization is likely to sacrifice the truth of your most profound philosophical tenets in order to realize less expensive epistemic gains elsewhere. (Think of what it might take to maximize knowledge of "the good."[3])

What does this have to do with my attempt to vindicate the inductive analysis of philosophical categories? Nothing at all if the philosophical concepts refer in a completely different way than the basic natural kind concepts. This book's working hypothesis is, however, the contrary thought: cognitively and semantically, philosophical concepts and natural kind concepts operate according to similar general principles. The same inductivist conceptual structure, the same source of cognitive significance in ordinary reasoning and in particular in inductive logic, the same general story (as you will see) about concept acquisition, and the same theory of reference—all are postulated in the hope of securing the objectivity and therefore the substantiality of philosophical knowledge.

But if objectivity must not be compromised, what to do about reflexivity? Which is it to be? The epistemic safety of the semantic sandbox or the dangers and rewards of the real world?

9.2 A Dispositional Theory of Reference

Objectivity and reflexivity can both be had in our representations of reality if we are willing to put off, for the time being, our claim to truth. This

---

[3] Williamson might not disagree. He invokes reflexivity in pursuit of a goal that is less ambitious than mine: he seeks only to show that there is a general tendency for beliefs to be true, so as to "[reassure] us that knowing is a natural state for believers" (2007, 277).

prospect of philosophical affluence through temperance is held out by the dispositional account of reference that I will now propose, based loosely on an idea suggested by Richard Boyd and supplemented by a verificationist ideal that can be followed back as far in history as philosophy goes.[4] (Johnson and Nado (2014) make a similar proposal.)

A dispositional theory of reference for kind terms—simple, but good enough for this book—might be formulated as follows:

> A term refers, at a given time, to a given kind just in case it is likely that, under ideal conditions, the term's users would apply the term to all and only instances of that kind.

Ideal conditions come in two parts: first, the reasoner's mind must be clear, capable, and capacious; second, the reasoner must be well informed.

Informed in what way? In my preferred take on dispositionalism, the thinker should have what I will call a total set of evidence; it is the categorizations they would make if they had this evidence that determine, at any given time, the reference of their terms. Evidence is the outcome of observation or experiment;[5] total evidence is a set of evidence sufficiently rich that no further evidence would substantially change your categorizing behavior. A total set of evidence is, unlike a set of all possible evidence, finite—or so I assume.[6,7] Evidence-based dispositionalism constitutes, in my view, a

---

[4] Boyd writes (apologizing for the "blurring of the use/mention distinction"): "Roughly, and for nondegenerate cases, a term t refers to a kind ... k just in case there exist causal mechanisms whose tendency is to bring it about, over time, that what is predicated of the term t will be approximately true of k" (Boyd 1988, 195). Another route to dispositionalism is by a generous application of a principle of charity "when trying to decide what the interpretee's terms refer to," as proposed by Jackman (2003, 161). To adequately survey philosophical verificationism or pragmatism would be a lifetime's undertaking; in this short footnote I will not pretend to try.

[5] Evidence, then, for the purposes of this study, is empirical evidence. It would, however, be easy enough to extend the notion of evidence to include other non-inferential sources of information about the world, such as direct introspection of the properties of conscious experience or certain forms of religious experience.

[6] This characterization of total evidence ought to be qualified: total evidence is a set of evidence sufficiently rich that no further set of *representative* evidence would substantially change the thinker's categorizing behavior. ("Representative" means representative of the kinds of facts that the thinker is consulting in determining category membership.) Sets of evidence gerrymandered to mislead the thinker, in other words, don't count.

[7] The notion of a total set of evidence presumes a certain large-scale uniformity to the universe, specific to the concept in question. What to do about a concept with respect to which no such

kind of verificationism: the reference of a term is entirely determined by the thinker's disposition, under certain circumstances, to apply that term; thus, there are no transcendent facts about reference. In what follows, however, I will use the more neutral term dispositionalism.

The dispositions that determine reference can be diagnosed, I think, by answering counterfactual questions: if I were to receive a total set of evidence, and to think clearly and so on, how would I apply such and such a term? Much of the work in setting up the dispositional account consists in stating more carefully what goes on in the counterfactual scenario in question.[8]

Evidence Acquisition

How is the total set of evidence delivered? Not linguistically, since it is precisely the truth conditions for language that are at stake. I therefore suggest that we imagine the thinker attached to what might be called an "evidence oracle": a machine that can simulate the experience that would result from performing any conceivable empirical observation or experiment. Ask the oracle what would happen if you sent such and such a satellite into space with such and such instruments, and it will respond, by way of virtual reality goggles, headphones, and so on, by recreating the experience of the operator in the control room receiving the satellite's data stream.

For the same reason that the oracle ought not to try to communicate linguistically with you, you ought not to try to talk to the oracle; in particular, you had better not be required to write down specifications of experiments and observations. The best solution to the resulting engineering challenge

---

uniformity exists, and so for which there is no total set of evidence? Dispositionalism denies that the concept has a determinate extension. So, I think, ought everyone else: the concept's *apparent* extension is permanently on the move. To assign any determinate extension, then, would be to postulate aspects of reference-fixing that do no work in the human cognitive economy. That is a coherent possibility, but it smacks more of faith than of philosophical reason. Or so the little verificationist in my philosophical conscience proclaims.

[8] The well-known objections to a counterfactual diagnosis of dispositions due to "finks" or "masks" ought also to be addressed, but they do not seem particularly pressing here, and in any case ought to be left to metaphysicians rather than semanticists. I assume that the relation between dispositions and counterfactual conditionals is close enough to make the following discussion useful.

is, I think, to imagine a Socratic oracle—an oracle that, without requiring (or even allowing) any input from your side, guides you through a set of observations and experiments that are sufficiently rich to be "total." After seeing what the oracle has to show you, then, your categorizing dispositions are settled, in the sense that no further evidence will materially affect the way that you apply your terms to the world. The Socratic oracle also sidesteps the difficulty raised by your not yourself knowing when your dispositions are settled. The oracle knows both the world and your mind; it ensures that the evidence it presents is sufficient to crystallize your categorical behavior once and for all.

There are infinitely many different experiments that might be performed in a world like ours; thus, there are presumably infinitely many total sets of evidence. It is conceivable that depending on which set is chosen, or the order in which it is presented, a term might end up being used in one of several different ways, or in other words, might end up being applied to one of several different possible extensions. Here the dispositional theory's probabilistic aspect—the "likely" in the formulation on p. 155—comes into play. Reference is determinate if one of these possible extensions is a much more probable endpoint of application than the others; otherwise, reference is indeterminate among the more likely extensions—as appears to be common in the early stages of science (Field 1973; Wilson 1982). Talk of what is probable implies some sort of measure over the different total sets of evidence and different orderings of those sets; however, I will not try here to nail down the source of the measure.[9]

It may in some cases be highly likely that, if enough evidence were to come in, a term would be abandoned—it would not be applied to any property or kind of stuff. In that case, the term is determinately empty, that is, it fails to refer.

Because dispositional probabilities can change, reference can change. A term might be highly likely to be applied in one way (upon receipt of a total evidence set), but then, because of some historical contingency—the term is commandeered by a headstrong and politically powerful researcher, say—it becomes highly likely under the same conditions to be applied in

---

[9] My inclination is to call upon the same sort of logical measure over possible worlds that I appeal to in Strevens (2011).

another (though typically related) way. The term "electron" provides a well-known example: originally introduced to quantify the minimal unit of electrical charge, it ended up instead referring to a kind of particle, the putative bearer of the charge (Fine 1975). Another example is "penguin," which originally referred to the Northern Hemisphere's great auk but was soon applied to, and in the course of the nineteenth century taken over by, the Southern Hemisphere birds that we now associate with that name.

The dynamics of reference change underline the fact that, although the dispositional account looks forward, it does not assign any role in determining reference to facts about the future. What fixes the reference of a term at a time is the character of a disposition existing at that same time. Were the universe to wink out of existence tomorrow, before anyone could accumulate a total evidence set and so begin to apply their concepts under ideal epistemic conditions, our concepts would nevertheless have always referred in virtue of dispositions that, though perfectly real, never had a chance to manifest themselves.

Case Judgments

Once the evidence is delivered, how are the decisions about cases made? Suppose that you are determining the reference of some language-user's term "swan" (and hence of their swan concept). Give them a total set of evidence; wait for them to digest it. Then what? You might imagine putting your subject in front of a conveyor belt, down which travels everything that might be categorized as a swan or as a non-swan. You record their decisions. To get a full measure of the term's referential reach—something closer to its intension than its extension—you also want to know how the term's user will treat non-actual but possible specimens, for some fairly broad conception of physical or biological or metaphysical possibility. That makes for quite a few objects on the conveyor belt. Can there be a fact of the matter about what your subject will do when confronted with uncountably many objects?

Rather than envisage a single counterfactual scenario in which a subject is asked to classify infinitely many objects, I suggest, we should imagine infinitely many scenarios in which the subject is asked to classify just one object. For each potential swan, then, imagine giving your subject a total

9.2. A DISPOSITIONAL THEORY OF REFERENCE

set of evidence and then asking of that one object: is it a swan? Collect the answers for infinitely many such scenarios, and you have your extension or intension. (Suppose for now that all classifications are made determinately and wholeheartedly; uncertainty and indeterminacy will be treated in Section 9.4.)

Idealized Reasoning and Categorizing Dispositions

To what extent, in interpreting the dispositionalist account, must we idealize the reasoning powers of the categorizer? We must assume that they do not mishandle the evidence. Because, in the inductivist view, what they do with the evidence is to build theories by inductive reasoning, we must therefore assume that categorizers are good inductive reasoners, meaning not only that they reason well (more or less in accordance with the canons of inductive logic, or if you like Bayesian epistemology) but also that they tend to see new theoretical possibilities where appropriate and abandon their original naive theory if or when the evidence begins to militate against it.[10]

I do not think that categorizers need be supposed to be perfect or ideal or "logically omniscient." They need only be like us on a good day, a day on which we live up to our human potential and no more. Although it may take them a while, with many false starts and steps, they will overcome in the end the limitations built into the normal human mind, reaching the same conclusions as a flawless reasoner.

Perhaps you are skeptical. Why think there is a fact of the matter as to how a rather ignorant and ordinary individual, however dogged and detail-oriented, would respond to something as mind-boggling as a total set of relevant evidence?[11]

Let me begin by observing that, in typical cases, a total set of evidence will not be so very vast. To determine what is and is not water, for example, you might have to attend only a week's intensive summer school. Here the

---

[10] Perhaps the evidence oracle can be fitted with a module that suggests (without endorsing) new theories where appropriate, relieving the categorizer of the burden of creation.

[11] On the question whether such sets of evidence exist, see note 7. The two questions are hardly distinct, given that the notion of total evidence is itself defined with reference to dispositional facts about a reasoner's handling of the evidence, but they are treated rather differently.

Socratic oracle may once more lend a hand, presenting only the evidence that is relevant to waterhood (or to put it more carefully, only the evidence that substantially impacts the way you apply your putative water concept). The counterfactual categorizer needs patience, but not superhuman resolve.

Further, the nonexistence of the categorical dispositions I require would be rather incredible, independently of dispositionalism, for reasons I now explain.

Take some term "K" that at time t refers determinately but concerning the extension of which its users are somewhat ignorant, and suppose that what dispositionalism needs is missing: among the term's users there is no fact of the matter as to how they will react to a total set of evidence. Suppose also that the world does not end any time soon, so that the users and their linguistic descendants eventually receive something like the total relevant evidence relating to "K." It might take only a few generations, or at least only a few centuries, as in the cases of "water," "gold," "electricity," and "heat."

If there is no determinate categorizing disposition, then there is no inevitable outcome to this process of theoretical development. After all, how could the end point of the historical process be any more determined, in advance, than the end point of the dispositionalist thought experiment? History is implementing the same controlled learning process that is implemented by the thought experiment, but with far more social, cultural, and political noise and many somewhat differently constituted reasoners rather than a single reasoner. If there is no fact of the matter as to what a single reasoner in epistemic isolation would do, it is surely not settled in advance what the multitude, weathering the storm of real life, will do with the same information. It follows that in real life, there are many forms that the "K" theory might happen to take by the time the total evidence comes in. Only a few of these forms, presumably, will pick out the extension that "K" has at the time t that the thought experiment notionally begins. Or to put it slightly differently: *By the time that the question of what counts as a "K" is finally settled by its users, it is highly unlikely that the set of things to which "K" is applied is identical to the set of things to which "K" referred at time t.*

That is a disturbing conclusion. Consider the possibilities. On the one hand, it might be that the extension of "K" does not change over the course of the learning process, so that it remains the same as it was at t. (I remind you that I am not assuming any particular theory of reference.) In that case, it is very likely that the set of things to which "K" is applied at the end of the process is not the set of things to which it refers. The acquisition of knowledge has led the term's users to apply it incorrectly, with no prospect of recovery.

On the other hand, it might be that the extension of "K" somehow does change, so that the things to which it is applied more or less mirror its extension. That is certainly a less appalling outcome. But it is still deeply unsettling. It implies that the users of "K" will never find out what they were talking about at time t, and indeed that they will never know that they were talking about something different at t than they were talking about later on.[12]

The same would be true for us, now. In matters where the reference of a term reaches beyond our current knowledge of the corresponding category, it would be not only possible but very probable that we would be forever cut off from the truth conditions of most of what we are currently saying—if, that is, the categorizing dispositions posited by dispositionalism are fictions. That doesn't make the dispositions real, but it does show that skepticism about their existence is itself a radical point of view.[13]

Contrast with Intensional Analysis

As I have framed it, the dispositional account might seem very close to the picture entertained by intensional analysts, who suppose that the mind represents, whether explicitly or tacitly, a criterion for category membership that will decide every case (while perhaps allowing for some borderline cases). I repudiated intensional analysis in Chapter 7 on the grounds of

---

[12] Here I am assuming, of course, that there is no independent way of determining the reference of "K" at t. That seems likely if conceptual inductivism is correct.

[13] The individualistic character of my brand of dispositionalism is suited to the lonely social ontology of the armchair. But dispositionalism could equally well be framed so as to turn on an entire linguistic community's categorizing dispositions. Some readers might find the existence of community dispositions easier to countenance. They would be sufficient for vindicating philosophical analysts' dependence on their starter beliefs provided that the ensuing analyses were sensitive to community opinion, as exemplified by the "positive" mode of experimental philosophy.

its incompatibility with conceptual inductivism. Does the dispositional account of reference let it in the back door?

No; there is a critical difference between the inductivism/dispositionalism package and the intensional analyst's psychology and semantics. The intensional analyst's criterion, however it is represented in the mind, has (and can be known to have) the logical status of a reference-fixing rule. It is stipulative: what it says about category membership not only goes, but is mandated to go. The self-given nature of the rule has two consequences: individual classifications can be known for sure to be correct, and the rule itself can be known for sure to be correct.

The inductivist/dispositionalist mind, by contrast, can never be sure of either individual classifications or the rule. Of course, if they could be sure that the dispositional account itself is correct, this would no longer be true; inductivism/dispositionalism would collapse into a "reference-first" variety of the intensional analyst's picture. But as remarked in Chapter 7, provided that the concept of reference is itself inductive, the nature of reference will not be known with certainty.

The inductivism/dispositionalism package is therefore incompatible not only with the intensional analyst's way of thinking but with any view in which "epistemic intensions" can be learned with certainty, such as Chalmers and Jackson's (2001) thesis that thinkers have the ability to "conclusively know" world-to-category conditionals of the sort discussed in Section 7.3.[14]

Compared to intensional analysis, dispositionalism also demands far less from the ordinary referrer in the way of counterfactual tendencies to classify. What matters for dispositionalism are dispositions, given current knowledge, to react to a set of total evidence. Intensional analysts care in addition about dispositions to react upon suspending or forgetting all current empirical knowledge, where such dispositions are putatively driven by a reference-fixing rule. Inductivists doubt the existence of such rules and (as I explained in my critique of Chalmers in Section 7.3) even putting the rules aside, they doubt the existence of the "blank slate" dispositions themselves.

---

[14] "If we possess a concept, then sufficient empirical information E enables us to conclusively identify the concept's extension, and we are in a position to do this whether the information in E is actual or hypothetical" (Chalmers and Jackson 2001, 325).

9.3 Reflexivity and the Vindication of Analysis

The dispositional theory permits anything and everything you believe about a category to be false, yet it is at root a reflexive theory: your present beliefs, even if they are entirely incorrect, are by the very nature of reference a starting point, if used logically, for finding the truth about the category to which they refer. (If the reference of a concept changes, as sometimes happens, the beliefs will also perforce have changed to a set that constitutes a proper starting point for inquiry into the newly referred-to category.) This homing tendency is sufficient—just barely—to vindicate an inductive inquiry based on your present beliefs. It does not give you reason to consider the beliefs to be reliable or justified, but it gives the beliefs enough of a hold on reality, however indirect, that you have good reason to go ahead and use them as the basis for an investigation of the categories they concern. Both reflexivity and objectivity (or at least objectivity's stand-in, the possibility of persistent and perhaps prodigious error) can therefore coexist, if the dispositional theory is correct.

The possibility of error extends as far as catastrophic reference failure. A term might be empty, if there is nothing to which we have the disposition to apply it in the light of total evidence. Dispositionalism is not so reflexive that it issues an ironclad guarantee of reference (nor should we ever have expected one). Does that undermine the vindication of analysis? No more than the possibility that theoretical terms in science might turn out to be empty undermines our justification for pursuing scientific inquiry—which is to say, not at all. One way to make progress in physics is to discover that there is no ether; likewise, one way to make progress in metaphysics is to discover that there is no causality. In both cases, dispositionalism gives you the reflexivity to ascertain, ultimately, referential indeterminacy and emptiness. (Whether the emptiness of philosophical terms such as "knowledge" or "causality" is as interesting a prospect as it sounds will be examined more carefully in Sections 13.3 and 14.7, when I turn to the question of the substantiality of philosophical analysis.)

The reflexivity needed to vindicate philosophical analysis can be found in the dispositionalist theory of reference, then; further, dispositionalism meshes harmoniously with the inductivist picture of the mind. That is of no help, however, unless the dispositional theory is correct. The only way we have to resolve such a question is, I think, philosophical analysis:

we must ask whether the dispositional theory satisfies our philosophical intuitions about the reference of category concepts and terms better than its rivals. Dispositionalism's ability to capture our case judgments about reference is my next topic. The aching methodological question posed when the foundations of analysis are founded in analysis itself will be confronted at the end of the chapter in Section 9.7.

9.4 Natural Kind Term Reference

My test bed for the dispositional theory is, as you might have foreseen, the basic natural kind concepts and terms. What I aim to show is that causal minimalism concerning the concepts and dispositionalism concerning reference combine to give the "intuitively correct" answers about reference in a range of familiar cases. What follows, then, is one part philosophy and one part cognitive psychology. The philosophy, in the form of dispositionalism, tells us that the key to reference lies in the categorizations we would make when given a set of "total evidence." The psychology applies the causal minimalist theory of basic natural kind concepts, along with some other psychological posits, to determine what case judgments normal humans would make in various circumstances upon receipt of relevant total evidence.

Let me begin with a not-so-tricky example for which much of the work has already been done.

A seventeenth-century English naturalist's ordinary beliefs about swans might, if the causal minimalist theory is correct, consist of core causal beliefs asserting, among other things, that something about swanhood causes white feathers, a red beak, waterborne mobility, and a tendency to trumpet. As we now know, several of these beliefs are not true of the genus *Cygnus*. Some swans are black; some are not trumpeters. But the concept nevertheless picks out the swans or some nearly equivalent group—so our considered judgments about reference strongly suggest. How can that be? My job is to show that the early modern English theory of swanhood will under conditions of sudden evidential omniscience be transmuted, by application of inductive thinking to the new data, into a theory that counts as a swan just those things that actually are swans, that is, just those things that fall into the genus *Cygnus* or some category close enough by.

History suggests that the task is not so daunting: biology began with something like the early modern theory of swanhood, or a collection of variants thereof, and upon the arrival of much pertinent information between then and now, fixed on a theory of swanhood—a rich set of ordinary beliefs about swans—that picks out the members of *Cygnus* with what we believe is a great degree of accuracy, meaning that we do not expect further information to greatly affect the category's boundaries. (We may of course discover new species of swan, but they will count as swans under the existing theory of swanhood, or so we confidently expect.)

It will be enlightening, however, to examine the fine-grained dynamics of the process of learning more about swans. How, exactly, do the facts we have observed or those we will observe shape the theory of swans, drawing it ever closer to a single, final theory of swanhood that delineates by its categorizations the exact boundaries of the extension of the past and future swan concept? While answering this question I aim to keep as close as possible to my inductivist precepts: the story should involve inductive reasoning about swans based on the influx of facts, but little or no reasoning about either the essential nature of swans or about the reference or any other semantic properties of the term "swan" and its corresponding concept.

Suppose, then, that you have a suite of core causal beliefs about swans typical of a seventeenth-century English naturalist (though perhaps not quite identical to anyone else's suite). As observable facts constituting a total set arrive in the manner stipulated by the dispositional theory of reference—as the putative Socratic evidence oracle leads you on a mute aural and visual journey of discovery through the animal kingdom—your theory of swans will change in several ways.

First and simplest, you will gain new core causal beliefs about swans. You must do some reasoning to get there, however. Nothing you observe will be directly attached to the term "swan." What you will be presented with is rather a lot of information about the insides, outsides, and behavior of a wide range of organisms. (They will not be characterized as organisms any more than they will be characterized as swans, but for expository purposes let me assume without argument that you recognize them as such.)

Some of these organisms you will classify, on the basis of their appearance and behavior—their lovely curved necks, their white feathers, their trumpeting and swimming—as swans. You do so in the usual causal minimalist way, using inductive reasoning, and more specifically causal reasoning, to infer that the best explanation of an organism's possessing such and such a complex of properties is its being a swan—since aside from swanhood there is no plausible causal explanation of there coinciding, in a cohesive clump of particles, that color, that shape, that kind of movement, that sound.

There are a bevy of organisms that you classify as swans, then. Thanks to your helpful evidence oracle, you will learn much about these organisms. You might learn, for example, that they all have double-chambered kidneys. (Again, I cheat a little by assuming that you get information about kidneys per se; in fact, you will get information about chunks of tissue that you will have to classify yourself as kidneys. Learning from the evidence alone, with no textbook to hand, is hard work.) You will naturally hypothesize that something about swans causes them to have double-chambered kidneys. In so doing, you gain an additional core causal belief about swans.

With this new information you will be able to refine (as you see it) your judgments of swanhood. A somewhat unusual-looking bird that you might have previously hesitated to call a swan might be classified as such with some confidence on the grounds of its internal physiology, once you have learned enough about the distinctive aspects of swan viscera.

Not every new belief about swans acquired from the oracle is a causal belief. As the evidence trickles in, you might come to believe that there are no swans in southern Africa or that Bulgarians like to keep swans as pets. This information, too, will play a role in the inductive thinking that drives your classifications: if you see, at a distance, a Bulgarian leading a large bird on a leash, you will in the light of your new ethnographic datum be more inclined than before to judge that it is a swan.

Sometimes beliefs will be subtracted rather than added. A flood of early information about Balkan pet proclivities might turn out to be, as more facts arrive, unrepresentative of the big picture; you will then abandon your belief about the Bulgarian affinity for swans.

And sometimes, of course, your core causal beliefs will have to be revised. At a certain point, what we would call the first black swan—the first Australasian *Cygnus atratus*—will come swimming down the data stream. You will react as Willem de Vlamingh and his crew did in 1696, seeing the first black swans floating down what they would later call the Swarte Swaene-Revier, now simply the Swan River (note 3 of Chapter 5). You believe that swans are white—that something about swans causes them to grow white plumage. But here are these birds that, color aside, display all the characteristic swan appearances and behaviors. How is this to be explained? Reasoning inductively like any good scientist, you infer that the appearances and behaviors are to be explained by the birds' swanhood, and that your causal hypothesis about swan color is too narrow. You amend your old core causal belief, substituting for the belief that swans are white the belief that swans are either white or black—or more probably, that something about the swans in Europe causes them to be white, and something about the swans in Australia causes them to be black.[15]

So your theory of swans changes and grows. As it does, it will come closer and closer to (though as the Bulgarians show, it may not converge monotonically on) a completely correct collection of facts about a property called swanhood. Some of these facts will be incidental (Zhivkov owned a pet swan). Some will be of principally nonbiological interest (Romanians have a great fear of swans). Such beliefs will play a role in classification only in certain cases and only when better information is unavailable.

But some will form a rich theory of the explanatory power of swanhood, relating that central property to various internal and external appearances and behaviors by way of core causal beliefs. It is this theory that the seventeenth-century naturalist will come to hold in the light of a total set of evidence, and so it is this theory's verdicts on swanhood that determine the reference of the naturalist's swan concept—which will, as a consequence, have the same reference as our modern concept attached to the same word.

Even at this supremely advanced epistemic stage—a stage where you have so much evidence that no further observation could change your mind—your categorizations are, I cannot emphasize strongly enough, still provisional. This for two reasons: they are based on ordinary beliefs, which are regarded

---

[15] The "swans in Europe" and the "swans in Australia" are in effect incipient species concepts. But let me not impose on you just yet the historically hard-won distinction between genus and species.

as in principle fallible, and they are in many cases inferred from those beliefs inductively. Like all the fruits of inductive inquiry, though they may be (and in many cases are) highly probable, they are never certain. In the case of a particularly atypical swan, your judgment may not even be confident. You may think that it is more likely than not a swan, but that the odds are at best 60/40.

In the dispositional theory of reference, this judgment is the final word on the specimen's swanhood. Does the dispositional theory rule, then, that it is definitely a swan? Not necessarily; dispositionalism is compatible with more than one answer to this question. The theory might declare a specimen to be a borderline case whenever your fully informed self is somewhat uncertain about its swanhood. Or it might count a specimen as definitely a swan if your probability for swanhood is one-half or higher, but otherwise as definitely not a swan.

My preference is for something closer to the former option, thus allowing for the existence of borderline cases in spite of the logical difficulties that they create, for two reasons.

First, as you will see in Section 15.3, the first option comes closer, I think, to our natural, pretheoretical beliefs about reference—a desideratum for the philosophical analyst.

Second, even if you opt for the maximally determinate second option, according to which the cutoff for definite swanhood is a subjective probability of one-half, some indeterminacy will creep into swanhood all the same. The source of this categorical fog is an indeterminacy in the fully informed subjective probabilities themselves: because the canons of inductive inference are somewhat permissive, and because the dispositional theory does not fix a unique total set of evidence or order of presentation, it seems quite unlikely that there is an exact fact of the matter about what subjective probability you would have for some problematic specimen's swanhood if a total set of evidence were to arrive in the next mail.[16] So there will be

---

[16] Traditional Bayesian conditionalization, executed flawlessly and with all possible hypotheses known in advance, does not allow the order of evidence to affect final judgments. Any deviation from this perfectionist ideal, however, opens the door. The order effects might be quite systematic if there are several different trajectories that your biological theorizing might take as the facts come

undecidable, hence borderline, cases after all (and, as with the first option, indeterminacy as to whether a case is borderline or not, an especially perplexing feature of vagueness). You might as well admit borderline cases cheerfully by taking the first option rather than, by taking the second option, try for determinacy and miserably fail.

˷

So much for straightforward swan studies. The rise of the causal-historical approach to basic natural kind term reference was, of course, accelerated by its exemplary treatment of some more recherché scenarios involving massive error, almost identical planets, and profound ignorance. Next on the agenda: to show how the dispositional account handles such cases gracefully, delivering the same answers as the causal-historical approach but by a quite different route. Concepts of philosophical categories continue to take a back seat, pending the development of my ideas about their inductive structure in Chapters 13 and 14.

9.5 Erroneous Beginnings

Schwanns: they are the pink-feathered, blue-beaked tromboners that you encountered on a foggy day at the beginning of the chapter. But these particular specimens turned out to be freaks, their appearance and behavior radically altered by exposure to a spill of printer's ink from the university press. Had the accident not occurred, they would have been, like almost every other one of their conspecifics, green-feathered and yellow-billed, with a plangent French horn call.

When I introduce the term "schwann" during my encounter with the abnormal specimens, I somehow succeed in attaching it to the schwann genus, though almost every belief I have about schwanns is false. How do I do it?

---

in, having as their endpoints several distinct biological theories of speciation, with each such theory drawing species boundaries slightly differently.

For the causal-historical theorist, the challenge is to show how my referential intention at the time I coin the term "schwann" can pick out the taxon despite the falsehood of my beliefs. The intention must, in effect, do an end run around the beliefs, getting its hooks into the schwanns by some other route. A natural suggestion, the centerpiece of the simple causal-historical approach described in Section 3.1, is that I introduce my term with the intention that it refer to all organisms with the same microstructure as the baptismal specimens. The facts about the ink-addled schwanns that determine the reference of "schwann," then, are not observable facts but rather some kind of molecular facts—patterns of DNA?—that are shared by all schwanns, normal and poisoned.

It is not so easy to explain precisely how I pick out *those* particular molecular facts, especially if my beliefs about what makes a schwann a schwann have nothing to do with microstructure (if I am a certain kind of Aristotelian, for example).

You might also worry how the causal-historical account can make sense of the following kind of case. Forget the ink; in the fens where I wander on that foggy day, all schwanns are normal: green feathers, yellow bills, horny when they call. The locals greatly admire the sleek looks of these beautiful animals, and build idols, remarkably true to life, in a number of places. Along my path I see no schwanns, but I do see, not too far away, one of the schwann idols. I mistake it for a real bird, and coin the term "schwann" to refer to such birds. Plausibly, my term as a consequence refers to the real, red-blooded birds—not to the idols, and not to nothing. Yet in that case the baptismal specimen has nothing microstructurally in common with the members of the term's extension. Indeed, the idols and the birds have nothing in common but their appearance.

I am here not, however, to argue against the causal-historical account, but to show you how the dispositional account deals with these same scenarios.

The ink spill first. How does dispositionalism secure the desired conclusion, our intuitive judgment that the term "schwann" refers from its inception both to the abnormal blue/pink birds and the normal but unobserved yellow/green birds? To answer this question, a dispositionalist asks: upon receipt of total evidence, would you come to believe that the yellow and green birds belong to the same category as the blue and pink birds?

9.5. ERRONEOUS BEGINNINGS

Here is one way in which you might find yourself making just such a discovery. You see yellow and green birds frolic in the spilled ink and later come to develop a blue and pink aspect. Then your thinking about your blue and pink baptismal specimens will follow roughly the same lines as your thinking about Keil's transformed raccoons. You reason, from their striking similarity to the birds you saw transformed before your eyes, that the baptismal specimens too owe their appearances to the causal equation: yellow/green + ink = blue/pink. Consequently, you change your beliefs about the normal coloring of schwanns, creating a unified theory of schwannhood that you believe covers both the baptismal and the normal specimens.

Or perhaps the similarity in overt body structure of the blue/pink and yellow/green birds is enough for you to reason as follows, upon seeing your first yellow/green specimens: these animals are so similar in their contours to the schwanns that I must conclude their aspect is best explained by my theory of schwanns. Thus, they are most likely schwanns. You will then amend your core causal beliefs as the Dutch sailors did after sighting Australia's black swans: where before you believed (or hypothesized) that something about schwanns causes them to grow blue bills, you now believe (or hypothesize) that something about schwanns causes them to grow bills that are either blue or yellow. (You do not yet suspect, of course, that the one complexion is a deformity chemically induced.)

Finally (but not exhaustively, because inductive reasoning can take infinitely many paths), you might find the blue/pink schwanns and yellow/green schwanns mating with one another, and even better you might find that they produce regular yellow/green progeny. Suppose you believe that, as a matter of biological fact, the great majority of sexually reproducing species breed successfully and only breed successfully with conspecifics. (Note that I've switched from folk genus to species; I won't comment further on this here.) Such a view can be represented using the core causal belief schema: for any species K, you believe that probably, something about Ks causes them to breed successfully only with each other. Such a belief is more than enough to give you strong inductive warrant for thinking that the two complexions of bird belong to the same species. (It also has the great advantage of allowing, as biologists would like to do without apparent conceptual incoherence, that there might be systematically successful outbreeding between some species, and even more important, that asexual organisms come packaged as species too.)

There are a number of routes, then, by which a good inductive reasoner might, in response to some limited but suggestive set of observable facts, come to classify the yellow/green birds, along with the blue/pink birds, as schwanns. The same inductive reasoner presented with all the evidence, as the dispositional theory of reference generously allows, will be surer still to put the two together. Thus from its inception, "schwann" refers to the yellow/green birds.

What about the case of the schwann idol? It is somewhat murkier, because the term "schwann," when newly coined, has several possible futures. It starts out embedded in a set of core causal beliefs—that something about "schwanns" causes green feathers, yellow bills, long curving necks, and all the other properties manifested by the idol—which would be perfectly accurate if "schwann" referred to schwanns. It is also embedded at this time, I presume, in some more general beliefs: that "schwanns" are birds, that they are animals, that they are flesh and blood, and so on.

Let me consider just the simplest course of later events. In this future history, you never realize your error, and you guilelessly apply your schwann concept to the birds themselves as you encounter them. When you come across an idol and recognize it for the inanimate object that it is, by contrast, you will not classify it as a schwann, since it evidently lacks many of the properties you attributed to schwanns from the very beginning, such as being animals, being made of organic matter, breeding among themselves, and so on.[17] That is a future where "schwann" comes, with more time and information, to be applied to the same things as it would have been if the baptismal specimen had been a real bird. If this future is the most probable, then right from the start your schwann concept refers to the bird taxon, even though you have not laid eyes on any of its members—simply in virtue of your living in an environment that overwhelmingly probabilifies the scenario described above.

---

[17] More subtly, but I think just as importantly, you will conclude that even your initial causal beliefs about schwann color are false of the idols, not because they are not the right color, but because you will see that there is no intrinsic property of the idols themselves that causes them to have that color. The core causal beliefs' presumption of intrinsicness is spelled out more explicitly in Section 12.5.

9.6 Identical Twins

There are other possible futures too; in some but not all of these "schwann" determinately refers to the birds; however, I leave the pursuit of this line of thought as an exercise for interested readers.

9.6 Identical Twins

Here are two categories, identical as far as you can tell—yet you have a term or concept that refers to one and not the other. What's going on? Let me discuss two famous cases presented by Putnam (1975): Twin Earth "water," and elms versus beeches.

Water

Suppose, Putnam imagines, there is somewhere else in the galaxy a planet much like Earth on which the seas are composed of a liquid with all the characteristic appearances and behaviors of water—transparency, potability, the power to dissolve salt, high surface tension—but which has a different molecular structure: rather than being made up of H_2O molecules, it is made up of X_yZ molecules. Take an English speaker in the early modern period, entirely ignorant of the molecular structure of water or indeed of the facts about molecular structure in general—Philoclea. Does Philoclea's term "water" refer only to the H_2O-containing stuff, or does its extension include also the X_yZ-containing stuff? We judge, says Putnam, that Philoclea's "water" refers solely to the former.

If the dispositional theory and causal minimalism are to work together to replicate this judgment, they must assign a high probability to Philoclea's constructing, upon inundation by a total set of evidence, a theory of "water" that excludes liquids that are largely X_yZ.

To show how causal minimalism predicts that she will very likely do so, I need to tell you more about the core causal beliefs. The belief that "Something about water causes it to be transparent" means roughly: samples of "water" robustly possess a property that plays a crucial role in the contextually salient mechanism for transparency (talking in a loose metalinguistic fashion to avoid begging questions about the reference of early modern "water").

There are two important aspects to this formulation. First, the "something about water" aspect of the core causal belief is to be interpreted, as

I explained in Section 6.1, as attributing to samples of water a (typically unknown) transparency-causing property—call it P.

Second, the core causal belief does not merely assert that there exists some causal mechanism by which the P-hood of specimens of water causes those specimens' transparency. It supposes that a particular causal mechanism has been made salient by the context, and it asserts that P causes transparency by way of *that very mechanism*.

Complementing the story is a criterion for individuating such causal mechanisms. This criterion will determine that H_2O causes transparency by a different mechanism than does X_yZ. More generally, it will determine that the appearances and behaviors of X_yZ and H_2O, though identical, are caused by different chemical mechanisms.

The core causal beliefs that make up Philoclea's theory of "water," then, collectively assert that "water" has its characteristic properties because of a certain suite of mechanisms. One such belief, for example, will say that "water" is transparent because of a particular mechanism, determined by context. Which mechanism? The mechanism at work in H_2O, or in X_yZ? I do not have to tell you much about contextual salience for you to foresee that it will be the former.

Philoclea's theory explains the transparency of "water," in short, by tacitly appealing to the chemistry of H_2O. The same goes for the other characteristic properties of "water." If Philoclea acquires a total set of evidence, she will therefore judge that H_2O-containing specimens' transparency, potability, and so on are explained by her "water" theory but that the same properties of X_yZ-containing specimens are not; consequently, she will classify H_2O-containing specimens but not X_yZ-containing specimens as "water."

The Twin Earth thought experiment is often taken to show that there is something indexical or contextual about the fixing of the reference of "water." My explanation, you can plainly see, accepts this lesson. What is contextual is the determination of the mechanisms, the putative consequences of which are described by the core causal beliefs. The mechanisms in turn help to determine what the theory is and is not true of, and thus they help to determine, according to the dispositional theory, what the theory's central term "water" picks out.

I had better answer some questions about contextual salience and the individuation of mechanisms.

The assumption that a core causal belief makes tacit reference to a contextually salient causal mechanism can be motivated quite independently of concerns about reference, namely, by the need to account for the "ceteris paribus" nature of the core causal beliefs and of our causal claims in general, both in science and in everyday life. The full story is given in Strevens (2012a). The short version is that when we say, "Ceteris paribus, F causes G," we make a tacit reference to a contextually salient mechanism M, concerning the workings of which we may know very little, with the "ceteris paribus" meaning "provided that the actual operation conditions for M hold." The reference to the mechanism accounts for the ability of the ceteris paribus hedge to qualify the causal hypothesis with conditions that its formulator may not yet be in a position to spell out.

How, then, did Philoclea manage to form beliefs that determinately pick out a mechanism concerning the workings of which she has not got a clue? More generally, how do the formulators of causal hypotheses pick out mechanisms whose operation conditions they do not fully grasp? It can be done using nothing more than a general criterion for mechanism individuation. You simply point to one or more baptismal samples, and you say, "I am making a claim about the consequences of the mechanisms at work in those samples" (thereby presupposing that the same mechanism is at work in the whole set). So, for example, you point to some substances that you consider to be water, and you say, in effect: "My core causal beliefs concern the mechanisms instantiated in these substances; I claim these mechanisms cause transparency" (or high surface tension, and so on).

Ordinary thinkers' criterion for individuating mechanisms is, I propose, supplied by the standard for the correctness of everyday causal explanations. When you judge that a stool's hitting a window is explanatorily relevant to its breaking, but that the stool-thrower's whoop of delight setting the window vibrating is not, you are judging that the one but not the other is a part of the causal mechanism that explains the breaking (Salmon 1997; Strevens 2008a). The rules for explanatory relevance are, then, sufficient for determining what counts as part of the causal mechanism producing any given event or phenomenon. Thus they are capable of individuating the mechanisms that generate the characteristic properties of various basic natural kinds. Philoclea, being a competent causal explainer of things, thereby has the power to attach mechanisms to her core causal beliefs without having the least grasp of their implementation. It is enough for her to say, "I intend my belief to specify the consequences of whatever mechanism explains transparency in this substance" (having in mind the stuff in the local lakes and rivers).[18]

For the sake of clarity, I have so far talked as though ordinary believers like Philoclea habitually make claims or express intentions about the contents of their core causal beliefs. That is unsatisfying, for two reasons. First, it is not in the spirit of conceptual inductivism, which does its best to avoid appealing to metalinguistic acts and beliefs in its explanations of everyday cognition. Second, it leaves the theory of reference dependent on a thinker's ability to "pick out a mechanism," which itself sounds like a rather referential activity. Let me see if I can do better.

In accordance with inductivism, I posit that a core causal belief's mechanism is determined by the thinker's ordinary beliefs about the category in question. What makes the H_2O mechanisms rather than the X_yZ mechanisms the subjects of Philoclea's "water" core causal beliefs are various other "water" beliefs, such as her belief that the stuff in the lakes and rivers around her is largely water (along with the fact that the mechanisms in the liquid satisfying those beliefs are H_2O-based). This is not because some special mechanism-fixing rule for core causal beliefs ascribes a leading role to the ordinary beliefs. No theory of mechanism fixing is necessary. All that's needed is dispositionalism along with the thinker's notion that their core causal beliefs concern particular mechanisms. Here's how it works. Philoclea gets her set of total evidence about water. To make her categorizations, she needs to figure out what's explained by her core causal beliefs about water. To do that, she needs to figure out what particular mechanism her core causal beliefs are about (since she starts out knowing that they are about some particular mechanism or other). She asks herself, "What mechanism am I trying to describe when I entertain these beliefs?" The ordinary beliefs about water will point her to the answer: she is trying to describe a mechanism that she supposes is at work in the stuff in the local lakes, rivers, and so on.[19]

---

[18] If such a stipulation goes wrong, note, it is not by misdescribing the consequences of the mechanism but by failing to pick out a mechanism altogether (Strevens 2012a, 669).

[19] The ordinary beliefs that matter are her current beliefs, that is, her beliefs at the time that we are using the dispositionalist theory to determine the reference of her term "water," not the beliefs

Even this train of thought is a little more self-conscious than that of the typical thinker, who upon getting a stream of total evidence will simply see that the mechanism at work in the stuff in the lakes and rivers, which because of their ordinary beliefs they take to be water, is built around H_2O. Throughout the process they think inductively; they need never ask, "What is my belief about?"[20]

The dispositional theory I have offered makes use, clearly, of many of the same materials as the causal-historical approach's theory of initial reference-fixing: reference goes by way of a contextually determined causal structure. The dispositional account is less immersed in the past, however, than the causal-historical account—it has no special use for causal chains leading back to baptisms—and unlike the simplest (and most clearly formulated) versions of the causal-historical account, it makes no use of referential intentions or other attempts to explicitly connect words to things.[21]

Three concluding remarks. First, let me emphasize that, in order to understand the workings of the dispositional theory of reference for basic natural kind concepts you need not only the thesis of causal minimalism but also some more specific theses that flesh out causal minimalism by giving a theory of the cognitive significance of beliefs of the form "Something

---

she entertained at the time she first formulated her core causal beliefs about water. This is not (in the first instance) to posit some rule about the semantics of core causal beliefs, but rather simply to point to the beliefs that Philoclea would naturally put to work to decide, at the time she notionally receives a total set of evidence, which mechanism her core causal beliefs are about.

[20] As an exercise, ask yourself what would happen if the stream of evidence arrived in this order: the characteristic properties of X_yZ and the causal details of the mechanisms thanks to which X_yZ possesses those properties, and only then the fact that the local stuff is made of H_2O, along with the chemistry of H_2O. I suggest that the thinker will at first come to believe that water is largely constituted of X_yZ. Upon discovering that the local stuff is not X_yZ, however, they will abandon this belief and identify H_2O as the causal factor underlying water's characteristic properties. Why would they not say, "Oh, so the stuff in the lakes and rivers is not water after all"? Why, that is, do they hold on to the belief that the stuff in the lakes and rivers is water while abandoning the belief that water is made of X_yZ? For purely inductive reasons, I suggest: they regard the former belief (and others like it) as far better established. Find a story where this is not the case—where the former beliefs are only tenuously held—and I suspect that your intuitions about the reference of "water" will begin to waver.

[21] The dispositional account is perhaps closest to Sterelny's (1983) version of a causal-historical account. In Sterelny's theory, the reference-determining microstructure of a natural kind term is picked out by its role in causing various observable characteristic properties of the kind; however, for Sterelny, unlike the dispositional causal minimalist, the relevant characteristic properties must

about Ks causes P." As I have said, these theses are independently motivated by the role that such beliefs play in scientific inquiry (see Strevens [2012b,2014] for "something about" and Strevens [2012a] for the contextually salient mechanism).

Second, if core causal beliefs and other causal claims work differently in a certain society, or subculture, or part of the world, then reference will also work differently there.

Third, in those places where core causal beliefs do have the structure I have described, believers will, on discovering the role of H_2O in the mechanisms explaining water's characteristic properties, come to regard as necessary for waterhood a substance's being constituted largely of H_2O. (What "largely" amounts to will depend on the details of the mechanisms.) It is possible using an inductive theory of natural kind concepts, then, to explain why some properties seem necessary, to the cognoscenti, for membership of a kind. Such necessities do not require definitions or referential intentions; it is enough that a concept's connection to the world is mediated by causal beliefs that go by way of particular mechanisms.

Trees

A similar kind of case, Putnam's elm and beech, calls for a different treatment. Philander, a botanically underinformed urbanite, has distinct concepts of elm trees and beech trees, so the story goes, but precisely the same beliefs concerning their observable and other properties—he really only knows that both are typical trees. How can the concepts have different extensions? Philander's beliefs about elms are not identical to his beliefs about beeches: he believes that one kind of tree goes by the English name "elm," the other by "beech." That small fact is enough to create a difference in the extension-determining disposition. If Philander were to learn more about trees, he would end up classifying some trees as elms and others as beeches—namely, just those trees that the experts classify under the same headings.

---

be fixed by stipulation at the time of baptism. And of course like any causal-historical account, and unlike the dispositional account, Sterelny's invokes causal chains leading back to a suitable baptism.

The process might unfold in several ways, depending on the order in which the evidence is revealed. Here is one possible course of events. First, Philander learns about trees. In the course of such learning, he discerns among many other biological categories two different kinds of tree, distinguished by their characteristic properties and the mechanisms that cause those properties, as described above in the case of water. He coins his own private Mentalese names *Fagus* and *Ulmus* for the two tree genera. Second, Philander learns some linguistic facts: *Ulmus* are what English speakers call elms, and *Fagus* are what English speakers call beeches. Being a good linguistic citizen, he proceeds to use his own English terms "elm" and "beech" in the same way. Thus he comes to apply "elm" to *Ulmus*, that is, to the elms, and "beech" to *Fagus*, that is, to the beeches. Had the facts arrived in a different order, I assert though I will not argue, the end result would have been the same. Thus as claimed, Philander's words, even when his botanical knowledge is in its initial parlous state, refer to elms and beeches.

The disposition in question depends in part, you will observe, on Philander's being a "good linguistic citizen." What does that mean? The simplest answer is that Philander supposes his "beech" to refer to the same things as other English speakers' word "beech"—or more generally, that he supposes himself to speak English. Thus, in order for Philander's "elm" and "beech" to refer to those two distinct genera, all that's needed is that he take himself to speak the same language as certain experts, a belief that is, though technically metalinguistic, of course ubiquitous and unremarkable, and presumably by anyone's lights essential to getting around in the human world.

9.7 Justifying Dispositionalism

The dispositional account of reference may be nice, but what reason is there to think that it's right? Through what philosophical methodology do we grasp its truth?

Philosophical analysis, I have said. Sitting in the armchair, then, or in the desk chair with psychology journals at hand, we consider what theory of reference best makes sense of what we take ourselves already to know about reference, both on a large scale and as manifested in judgments about what refers to what in particular cases—Twin Earth, blue gold, Gödel, Clyde the moose, and all the rest. The greater part of this chapter consists of such analysis; finding a match between intuitions and predictions, it gives dispositionalism a clean bill of health, opening the doors to a vindication of philosophical analysis founded on referential reflexivity.

But then how to vindicate the philosophical analysis of reference itself? What gives us prior warrant to suppose that our untutored ordinary beliefs can be relied on to point the way to the truth about reference?

There can be no independent answer to this question, thus no foundationalist argument for the dispositional theory of reference or for philosophical analysis itself. Apostles of the inductive way are quite used to this circular or bootstrapping aspect to the ultimate epistemology of almost everything. We cannot justify inductive reasoning without recourse to induction, and in the same way, we cannot justify philosophical analysis without recourse to armchair philosophy. That is a consequence of the way our minds are built. An a priori foundation for philosophical or other substantive thinking requires an extraordinary belief, that is, a belief whose warrant can be apprehended directly through its stipulativity or some other supplier of self-evidence. We have been allotted only ordinary beliefs—and no ordinary belief can crown itself or any other ordinary belief with extraordinariness.[22]

In that respect the "vindication" promised in the title to this book is compromised in the same way that all the great foundational epistemological projects are compromised. The best I can do for the seeker of ultimate grounding is to say in a mildly Humean vein: fall back on your instincts, and you will find yourself contemplating an attractive picture. Thanks to a modicum of uniformity in the way the world works, inductive thinking provides rich knowledge of the empirical world (and more); thanks to a modicum of reflexivity, armchair reflection provides rich knowledge of the philosophical world (and more).

But I will not stop there. Although the dispositional account is in its architecture quite different from other views of reference, it is in its consequences quite similar: given some plausible assumptions, several different

---

[22] Genuine a priori knowledge may be unattainable regardless of the way our minds are built. But here the psychological premise is both sufficient and illuminating.

views of reference imply what the dispositional account presumes by its very nature, that after enough evidence comes in, our judgments of category membership will be exactly correct.

Let me confine my argument for this claim to categories of physical objects. Assume that all facts about the material world are susceptible to empirical discovery, so that a total set of evidence will settle—not conclusively, but as accurately as you like—any question about the constitution of material things.

Now consider the simple descriptivist account of reference, on which every concept is associated with a description that directly determines its extension. What happens once a total set of evidence arrives? Provided that the description and its apodictic role are clear to the categorizer, they will see which material things do and which do not satisfy the description, and so they will make correct category membership judgments in every case.

Or consider the simple version of the causal-historical approach in which the reference of a natural kind term is fixed by the user's intention that the term refer to everything with the same microstructure as a baptismal specimen. Given sufficient evidence, the user will discern both the original sample's microstructure and which other things have that same microstructure. Classifying in accordance with their original intention, they will make no mistakes.

A causal-historical theory that appeals to "reference magnetism" to steer reference toward some extensions at the expense of others does not lead so directly to a harmony of categorization and category. If it is possible to uncover the workings of this occult semantic force, however, the categorizer will get there in the end.[23] Were reference to depend in part on what is "natural," for example (Lewis 1984), then provided that we can through observation and reflection learn what is natural and what is not, the fully informed categorizer will have the resources that they need to apply their terms correctly.

Both the descriptivist and the causal-historical views of reference, then, imply (given the additional assumptions about the transparency of the description and the referential intention, the empirical accessibility of

---

[23] As explained in the online-only appendix 2 to Chalmers (2012): "Reference Magnets and the Grounds of Intentionality," at http://consc.net/ctw/appendix2.pdf.

material reality, and so on) what you might call the dispositional doctrine of reference:

> A term refers, at a given time, to a given kind just in case it is likely that, under ideal conditions, the term's users would apply the term to all and only instances of that kind.

(As before, for simplicity's sake inconclusive judgments are in this formulation ignored.)

I cut and pasted that sentence from the statement of the dispositional account at the beginning of Section 9.2. The dispositional account and the dispositional doctrine differ only, then, in that the account but not the doctrine takes the sentence (or some more sophisticated variant) as stating the essential nature of reference, rather than as stating merely one of the essential nature's consequences.

The dispositional doctrine provides all of the reflexivity of the dispositional account; thus, it vindicates the philosophical analyst's reliance, as a starting point, on their initial philosophical beliefs—even in those cases where the reference of a philosophical concept is unstable or fails altogether (Section 9.3).[24] This suggests an alternative approach to the one taken in this chapter: rather than arguing for a particular theory of reference, argue for the dispositional doctrine, and the sweet reflexivity it brings, on the grounds that it is implied by all plausible candidates for the true theory of reference.

There is, of course, something strategically savvy about calling on the weakest possible premise, and I will in the following pages occasionally remind you that only the doctrine, and not the account, is needed for my purposes. (The exceptions—the places where I need the account and not merely the doctrine—are in Chapters 11 and 15.)

But I do not think that the suggested argument for the doctrine, that it is implied by all the better candidates for a philosophical theory of refer-

---

[24] It also provides an effective way of evading Mallon et al.'s (2009, 332) argument against attempts to "derive philosophically significant conclusions from theories of reference," on the grounds that different cultures have different notions of reference. If everyone's notion implies the dispositional doctrine, then it appears safe to derive conclusions from the doctrine, if not from more parochial particulars.

9.7. JUSTIFYING DISPOSITIONALISM

ence, has much force. The reason is a certain "interesting" feature of the dispositional doctrine.

Suppose that we attempt to determine the nature of reference by philosophical analysis, as I have been doing in this chapter. The principal source of evidence for and against various hypotheses about the true or essential nature of reference is our case judgments, that is, our considered opinions about what terms or concepts refer to what. In the case of category terms or concepts, the case judgments presumably conform to the reference-oriented version of the T-schema for kind terms:

x belongs to the extension of "K" just in case x is a K.

Thus our judgments about the reference of our terms will exactly correspond to our judgments about membership of the corresponding categories.

In that case, however, we will consider a theory of reference completely successful in capturing our case judgments about our terms just in case it matches our categorizations under those terms. Further, we will consider our judgments about reference to be final just when our categorizations are final, that is, after receiving a total set of evidence. At that point we will regard a theory of reference as fully adequate to our "intuitions" just in case it satisfies the dispositional doctrine (still restricting the claim to our own terms, though the argument could be extended using a simple and plausible hypothesis about the way in which we make judgments about the reference of other people's concepts and terms). It is in the very nature of philosophical analysis, then, that it will tend to settle on theories of reference that imply the dispositional doctrine. Somehow, this observation does not fortify my confidence in the doctrine as a foundation for analysis. But I suppose that I am in any case a true enough believer to press forward.

CHAPTER 10

The Travails of Analysis

10.1 Philosophical Analysis Is Doomed

The short history of twentieth-century philosophical analysis is one of failure. As summarized succinctly by two well-known writers:

> *No* commonsense concept that has been studied has turned out to be analyzable into a set of necessary and sufficient conditions (Stich 1992, 250).

> The pursuit of analyses is a degenerating research programme (Williamson 2000, 31).

Does this wretched record mean that philosophical analysis is hopeless? That you should put down this book—permanently—and take a long walk?

The next two chapters confront the question of failure head-on by examining a range of explanations, including those given by Stich and Williamson, for the stalling and breakdown of the program of analysis. I assume that philosophical concepts are inductive, which turns out to undermine certain explanations of failure and to provide an account of instability and diversity in case judgments that saves philosophical analysis from the skeptical attack mounted by some experimental philosophers. That is the good news for philosophical analysis.

The bad news is that inductivism inspires a new explanation for the failure of analysis that, if correct, suggests that many important philosophical categories will indeed be unanalyzable. Yet among the ruins, two reasons to stay in the armchair remain. The same inductivist explanation shows, first, how some important categories may be successfully analyzed, and second, how the analyst's primary technique—the method of cases—can discern truths about the philosophical categories that, even if they fall short of a complete analysis, are sufficiently substantive to constitute philosophical knowledge.

How could the analysis of a category—the attempt to find the category's essential nature by testing hypotheses about that nature against judgments about cases—possibly go wrong?

Here are four paths to disillusionment and despair:

1. You cannot find a theory of your category's essential nature that replicates the judgments we humans make about category membership; any candidate theory gets some such judgments wrong (or perhaps fails to pronounce at all on some important cases).
2. You have a theory that more or less replicates the judgments, but it is disappointingly messy—full of "epicycles," an "ad hoc sprawl" that suggests an arbitrary, uninteresting kludge; an intellectual deposit of psychological, social, and historical detritus, rather than a category of supreme philosophical importance.[1]
3. You have more than one theory that replicates the judgments, and no way to choose among them.
4. You have more than one theory that replicates the judgments in everyday cases. What would decide among them are their diverging pronouncements about certain outré cases, but philosophers' judgments differ or go indeterminate on these cases.

We philosophical analysts are quite familiar with the first and second predicaments. Robert Shope's *The Analysis of Knowing* (1983) is often cited as a miserable litany of repeated failed attempts to provide necessary and sufficient conditions for "*s* knows that *p*." The literature on singular causation has its own version of the story, in which promising attempts to analyze

---

[1] Williamson (2000, 31) on the concept of knowledge: "The difference between knowing and not knowing is very important to us . . . This importance would be hard to understand if the concept *knows* were the more or less ad hoc sprawl that analyses have had to become; why should we care so much about *that*?"

"c was a cause of e" bog down in a similar way, piling on additional clauses and complexity to neutralize an apparently unending stream of counterexamples.

The happy misfortune of having too many good theories—the third predicament—is rather more rare, though I will give an example of something in the vicinity in Section 10.3 below.

The final problem, of encountering "clashes of intuition" just when it matters most to have a clear case judgment, is also familiar to long-time denizens of the armchair, but has been made newly salient by recent work in experimental philosophy suggesting, as explained in Section 4.4, that there may be many more clashes than we professional philosophers suppose. Gettier cases, Kripkean Gödel judgments, water and Twin Earth: in all of these famous cases, the canonical intuition—the case judgment anointed by professional philosophers as correct—is found not to be shared by other cultural groups or individuals, or to be rather unstable even in those individuals who tend to go along with the received philosophical wisdom.

The aim of this and the next chapter is, as I have said, to ask how the hypothesis of conceptual inductivism might explain, or even give us reason to expect, the four varieties of failure just described. My method, as throughout the book, is indirect: I will take as my principal topic, at first, the analysis not of the philosophical categories but of the basic natural kinds.

To analyze a category such as water is to determine its essential nature. As I observed in Section 1.4, such an investigation is heavily empirically informed. But it cannot be concluded entirely in the laboratory; it has a significant armchair phase. Imagine that all the relevant chemical information is in: we have become experts in H_2O and its properties, both pure and when mixed with other substances. These facts do not, in themselves, tell us when something counts as a specimen of water. What percentage of H_2O must a liquid contain to be considered water? Is it, indeed, a matter of percentage alone? To answer these questions, the metaphysician of water considers various possible mixtures and asks, "Does this count as water?" And they reject hypotheses about water's essential nature that do not recapitulate their judgments. They proceed, in other words, according to the method of cases.

It is striking that such attempts to spell out the essential nature of basic natural kinds run into many problems familiar from philosophical analysis: never-ending counterexamples, epicycles within epicycles, intuitions that fog over in just the places where progress demands clarity. Perhaps the explanation for the agonies of philosophical analysis and that of the difficulty of finding essential natures for basic natural kinds have a common foundation in the inductive nature of the corresponding concepts? I propose to act on that suspicion, using the example of the natural kinds as a stepping stone to understanding the vexations of analysis in general.

Let me emphasize that the objects of analysis in what follows are our pretheoretical basic natural kinds: the category of swans picked out by both Locke's and my own term "swan"; the category of water picked out by both Philoclea's and my own term "water"; the category of beeches picked out by both Philander's and my own term "beech." Science has overlaid on top of this old classification scheme new and more finely grained classification schemes of its own: the Linnaean system dispenses with "folk genera," replacing them with genus and species;[2] the chemists meanwhile distinguish isotopes, elements, and compounds, not to mention elemental metals and their crystalline forms, pure and alloyed. I am putting all of that aside, giving an account of the everyday categories we've inherited from our ancestors, not the new categories we've created to sit alongside them.

10.2 Water

What is water? What property shared by the stuff in the sea, the stuff coming out of faucets, the stuff in this beaker, makes them all specimens of water, in the ordinary sense of the word?

---

[2] As I noted in Chapter 5, there is a remarkable correspondence between the Linnaean taxa and the lower-level pretheoretical biological categories in many cultures around the world (Berlin et al. 1974; Atran 1990). This suggests to me, as it has suggested to the anthropologists who discovered it, that scientific and naive taxonomy in biology (and I would add chemistry) are very closely connected, and indeed, that some aspects of scientific taxonomy are little more than naive taxonomy perpetuated in formal guise. For the sake of the argument, I make nothing of this hypothesis in these pages, supposing that scientific and naive classification schemes are in principle distinct, even if they seldom cross-classify.

10.2. WATER

It is a commonplace that water is H_2O, but as everyone who has tried their hand at the water question knows, this platitude is very far from comprising an adequate philosophical analysis of the nature of water. There are a number of arguments to this effect in the literature, not all entirely consistent with one another; what follows is the line of argument that I find most convincing, a pastiche of various considerations borrowed from other thinkers, such as Malt (1994), Chomsky (1995), and LaPorte (1996).

First, water is not pure H_2O. Many substances that we correctly believe to be kinds of water—tap water being an obvious example—contain much else besides H_2O molecules.

You might reply that when we call such liquids "water," we are referring only to the part of them that is composed of H_2O molecules, just as when we call the figures we see moving in the street human, we are talking about the animals but not their clothes. If we were more careful, we would say that tap water is not entirely water: it is a mix of water and chlorine and salts and various other stuff. But this cannot be right. Pure H_2O is not conductive: it is the dissolved electrolytes found in any sample of water outside a lab that give it its conductivity. Yet we say (or the chemists tell us) that water is generally conductive. When we talk about water, then, we are talking about the whole liquid, not just the H_2O.

At the very least, it seems that water must contain some H_2O—perhaps a large amount. It is natural to suggest, then, that a substance's waterhood consists in its having a sufficiently high percentage of H_2O. There might be a precise cutoff point, or (more plausibly) there might be an intervening range within which waterhood is indeterminate.

This proposal cannot be right either, however, for reasons suggested by Malt (1994): some kinds of water have considerably lower percentages of H_2O than some kinds of non-water. A cup of moderately strong coffee, for example, is about 98.8 percent H_2O; seawater is about 96.5 percent H_2O (both percentages by weight). Coffee is not a kind of water, whereas seawater, despite its impurities, is.[3]

---

[3] Malt's experiment, invoked as an argument against psychological essentialism in Section 5.4, examined subjects' beliefs about the percentage of H_2O in various liquids; since we are doing metaphysics here, I cite the actual facts about the percentages instead.

What makes the difference, it would seem, is the effect of the added ingredients on the properties of the liquid. Something about flavor and perhaps color appears to matter, since coffee and seawater do not differ much in their other properties. If there is a metaphysics of waterhood, then, it must take flavor among other things into account if it is to explain why coffee is not a kind of water.

The connection between flavor and waterhood, however, is far from straightforward. You cannot exclude all coffee-flavored liquids from the category of water: the water I use to wash out my coffee jar acquires a coffee-ish taint, but in so doing, it does not relinquish its status as a kind of water. You will need to find some cutoff point at which the flavor is too strong for the liquid to remain water. Yet there is coffee on offer in some places that is barely coffee flavored at all. Could it be that part of what it is to be water is not to be coffee, and part of what it is to be coffee is to be brewed with a certain intention? If the kitchen hand is, however ineptly, attempting to make coffee, then a certain weak mix may count as coffee; if the same weak mix is produced by a washing-up incident, then it may count not as coffee but as coffee-tainted water. In that case, the absence of a certain intention will find its way—bizarrely, epicyclically, ad hocly—into the metaphysics of water.

What is already an uphill battle gets more formidable still because the case judgments that distinguish different putative solutions to the problem are themselves hazy or controversial. In my own informal investigation of other people's judgments of waterhood, for example, a minority are inclined to count tea and coffee as "kinds of water." On tears, my informants are split: half say that they are a kind of water, half that they are not.

To specify the essential nature of our naive water category is therefore, in spite of its central role in our and our ancestors' thinking, difficult in many of the same ways that philosophical analysis has turned out to be difficult.

10.3 Swans

What do all swans have in common, in virtue of which they are swans?

A common answer is, to put it crudely, "swan DNA." The DNA molecule appeals as a locus of biological essence for two reasons. First, it is the genetic material, the material of inheritance. Second, it is what is responsible for the zygote's growing into the complete organism, with its distinctive color, beak, behavior, and so on. Both of these notions—particularly the latter—are oversimplifications. But for the sake of the argument, let me ignore the concerns of the biological sophisticate and suppose that there is a way of characterizing a region of the space of all possible genotypes such that an organism falls within the region just in case it is a swan. Say that such an organism has the *genetic signature* of swanhood. You might think that the essence of swanhood is its genetic signature. An organism is a swan, then, in virtue of, and solely in virtue of, its having the signature.

This molecular metaphysics of the swan and other folk genera appears to be quite pervasive among nonphilosophers. A corresponding approach to the Linnaean categories of species and genus is not at all popular among biologists or philosophers of biology, however.[4] The Linnaean categories are not identical to the folk categories, and so need not have similar essential natures—yet promising strategies on the scientific side do seem to work rather well on the prescientific side (see also note 2). Let me therefore borrow some ideas from analysts of the Linnaean categories, using them to sketch possible analyses of the swan folk genus to rival the molecular account.

A phylogenetic view of swanhood focuses on evolutionary lineage. The swans are a piece of the evolutionary tree, a *cutting*, as it were. What it is to be a swan is simply to fall into the relevant cutting, the relevant segment of the tree. (In biological taxonomy, there are several rival rules for determining the allowed geometry of such a cutting, of differing levels of strictness.)

Another view of swanhood emphasizes the functional connectedness of organisms, and in particular their ability to breed with one another. According to Mayr's "biological species concept," a species is a group of natural populations that actually or potentially breed together. (Mayr is motivated by a robust realism about species that rejects any metaphysics that draws what appear to be arbitrary boundaries in the space of genotypes or in the phylogenetic tree; his aim is above all to explain why the boundaries appear where they do.) Mayr's species concept cannot be applied directly to the swan folk genus, because the folk genus encompasses a number of

---

[4] Among philosophers, perhaps Boyd (1999) comes closest to endorsing a sophisticated version of the genetic signature view. His concern is not with genes per se, but he puts a high premium on the properties of an organism that maintain its bodily and ecological integrity.

different species that do not have a natural tendency to interbreed (although in some cases interbreeding is physically possible, as between the mute swans of Europe and the black swans of Australasia). An indirect application is, however, possible: use Mayr's criterion to delineate the various "folk species" of swan, and then unite these into the swan folk genus on the grounds of both their evolutionary relationship and their morphological and ecological similarity. Mayr (1981) calls this (when applied to Linnaean taxa) the "evolutionary" approach to taxonomy.[5]

How might we choose between the molecular, phylogenetic, and evolutionary approaches to understanding what it is to be a member of the swan folk genus—the category picked out by the ordinary English word "swan"? As is well known in the philosophical community, lightning strikes in wetlands can occasionally create molecular replicas of swans, the electrical energy by utter chance assembling from the raw organic material of the swamp something with the same morphology and the same DNA as normal swans. Are these cygnid swamp progeny real swans? According to a theory in which the essential nature of swanhood is nothing over and above a certain genetic signature, the answer is yes. According to a theory that requires the swans to make up a cutting of the evolutionary tree of life on earth, the answer is no. So which is it?

Judgments conflict. Some say that the swamp creature is a swan, most that it is not. (Almost all biologists would say that it is not, but their views are disqualified, because they are presumed to be applying the Linnaean, not the naive, taxonomy.) Perhaps those who classify the swamp creature as a swan are making an inferential error, in which case further reflection of sufficient quality and intensity will persuade them to retract their judgment. Suppose for the sake of the argument—though I suspect far too optimistically—that they do so; everyone eventually comes to agree, then, that swamp cygnids are no swans. Does that rule out the "genetic signature" theory of swanhood? Not entirely; we can have a genetic signature

---

[5] Yet another metaphysical thesis about the nature of species holds that they are not kinds but individuals (Hull 1978). The view is somewhat orthogonal to the rest: regardless of whether the swans are a kind or an individual, it is possible to ask what makes something a swan and to give as an answer one of the three views above—though of course kind theorists will interpret the answer as a criterion for category membership, while individualists see it as a criterion for a certain type of parthood. For expository purposes, then, I put the question of individuality aside.

theory that adds as a necessary condition for swanhood reproductive connectedness to the other swans. That leaves us with three options again—the genetic, the phylogenetic, and Mayr's evolutionary approach—though the first has moved a little closer to the other two.

To distinguish among them, you might look to see what they have to say about the temporal boundary of swanhood, that is, about the question of when the evolutionary ancestors of swans became swans proper (presumably via a succession of borderline cases of swanhood). Here you might hope that, say, genetic and ecological criteria would come apart. But on this question of priority—which of the swan ancestors were the first swans?—all three accounts of swanhood go mushy, revealing themselves to be templates for specifying swanhood's essential nature rather than fully specified theories.

To say that swanhood is a matter of DNA, for example, does not determine any particular zone in the genetic transformation from swan ancestor to swan as that during which the new species appeared. The same is true for the phylogenetic approach to swanhood. Mayr's criterion is a little more helpful, if there was a stretch of time during which before and after reproductive compatibility changed rather abruptly. Presumably analysts of swanhood will fill out the details of the genetic and phylogenetic templates to accord with intuitive judgments of swanhood, insofar as consistent judgments of this sort can be elicited—but then the fully specified theories that result will agree on all clear cases.

What confronts us, it seems, is a genuine case of theoretical underdetermination. The evidence—the totality of human judgments about what is and is not a swan—is correctly predicted by more than one plausible theory about the essential nature of swanhood, about what makes things swans. Superempirical considerations such as simplicity, or harmony with overarching theoretical maxims such as the importance of descent with modification, might help to decide between the options, but then again, they might not. (There is no doubt that the tree of life should matter to the theoretical categories of evolutionary biology, but it is less obvious that it should dominate the metaphysics of naive categories, such as folk genera, given their more practical function.) If there is a fact of the matter as to the nature of the swan-making property, it would appear to be closed off to us forever.

In the case of water, we cannot find a plausible account of the category's essential nature that replicates our expert judgments about individual specimens' waterhood. In the case of swans, we can find too many plausible (if as yet incomplete) accounts of the category's essential nature that replicate our expert classifications.

To accommodate some kinds of judgments, clauses that appear to be epicyclic and ad hoc—the need for a certain intention to be missing, for example, to distinguish strong washing-up water from weak coffee—must be added to the analysis. Further, at just that point where a clear judgment about a certain case, such as the swamp swan, promises to discriminate between otherwise equally attractive analyses, some participants' intuitions falter and fade. It is all rather familiar to an experienced philosophical analyst. This is how the searches for the essential nature of knowledge, of causes, of moral responsibility, find themselves in each instance stranded a few miles short of their destinations.

Approaches to understanding the travails of analysis might take one of three routes. First, there are those that contend that the categories to be analyzed have no essential natures, an explanatory strategy to be examined in the next chapter. Second, there are those that allow the existence of essential natures, but contend that knowledge of these natures is in some sense disappointing: perhaps the natures are too complex, too arbitrary, or insufficiently objective. This strategy will be considered in Chapters 12, 13, and 14, where I take on the problem of substance. Third, there are those that admit the existence of interesting essential natures, but that posit formidable impediments to attaining knowledge of such natures. Several strategies of this sort are examined in the remainder of the present chapter.

10.4 Conceptual Complexity

"In retrospect," writes Sider (2011, 116–117), "we should never have expected [philosophical analysis] to succeed." He continues:

> Why should there be any simple definitions, preserving intuitive or cognitive significance, of any of our words in any other terms? Words aren't generally

introduced as definitional equivalents of pre-existing phrases, and even then they subsequently take on semantic lives of their own. Current meaning derives from a long, complex history of use, which would seem unlikely to result in neat equivalences.

Sider's explanation for the difficulty of analysis, then, goes something like this. The complexity of a concept will likely mean complexity in any halfway adequate theory of the corresponding category's essential nature. The typical philosophical concept is extremely convoluted, thanks to a nondefinitional introduction and "a long, complex history of use." Thus the essential natures of philosophical categories, should they exist at all, will tend to be extraordinarily difficult to learn. (Chalmers [2012, §§1.2, 8.2] has ventured a similar story about the frustrations of philosophical analysts.)

A conceptual inductivist will be more than ready to agree with Sider that our philosophical concepts are not acquired by framing definitions, and that as history unfolds, they tend to accumulate rather than shedding cognitive complexity, as the theories attached to the concepts grow. Does it follow from this proliferation of complexity in conceptual structure that essential natures will become equally tangled and tortuous?

That conclusion might not be unreasonable if, say, the Ramsey-Lewis view of reference, according to which a concept's extension is fixed by the sum total of concept-involving "platitudes," were correct. With the passing of time, a concept's set of platitudes will tend to increase in size. Because the set is a reference-determiner rather than an explicit specification of an essential nature, it does not follow that the category picked out by the concept has a nature of analysis-defying intricacy. But it is reasonable to suppose that, as the platitudes proliferate, the constraints that jointly determine the nature become ever more difficult to satisfy in a simple way, so that our categories become more convoluted the more opinions we accumulate and the more knowledge we gain. All we frustrated analysts could do, in that eventuality, is to dream about those sweet, raw, Edenic days before our own meaning-creating acts alienated us from the categories about which we talk.[6]

---

[6] Since according to Sider's picture our categories are always on the move, the philosophical kinds that we care about today are strictly different from the kinds that a philosophical Eve might have cared about in the garden. Our tree of knowledge bears very different fruit . . . .

If conceptual inductivism is correct, however, then the preceding paragraph's story is entirely off the mark. Although there is no question that the cognitive structure of (for example) basic natural kind concepts complexifies over time, getting larger and more elaborate as humanity learns and ages, this need not, and I think typically does not, result in a change of reference. A natural kind concept becomes richer because we acquire more beliefs about the kind. If it is the same kind, then it has the same essential nature as it did upon the concept's introduction; history has not magnified its complexity.[7]

Indeed, as we come to know more about a kind, and so as our concept of that kind becomes more elaborate and sophisticated, our judgments about category membership become more extensive and more accurate, and so it becomes *easier* to do analysis. What accrues with time is not the complexity of a category's essential nature, but rather the stock of clues we have to decode the essential nature. From a causal minimalist's perspective conceptual complexity cannot, then, explain the difficulties we face in analyzing water or swanhood. Nor can it explain, if philosophical concepts are inductive, the quandaries of philosophical analysis.

Inductivism, paired with the dispositional account of reference, does allow the categories picked out by concepts or words to change. So, you will recall, "penguin" might start out referring to the great auk and end up referring to a family of birds in the Southern Hemisphere. In such a case, the change in reference hardly impedes analysis, but are there other cases where the new category is more complex and so more difficult to analyze? I think that in general the opposite is true: when reference changes, the new category is usually better aligned with objective markers and so easier to analyze.

Consider, for example, the word "insect." Upon its introduction in the seventeenth century, the term was applied, according to the *New Oxford American Dictionary*, to "any small cold-blooded creature with a segmented body"—so not only insects in the modern sense, but millipedes, spiders, and so on. You might suppose that at that time, the naive theory of insects

---

[7] In this respect, the inductivist view is close to the classical view in which a simple, fixed definition is the semantic sun around which the beliefs and other cognitive debris accumulate over time.

was too spare to determine, by way of the dispositional theory of reference, a definite class of animals as the term's extension. Too many future histories were possible. The term might have remained an informal term picking out any animal satisfying roughly the above description, rather like "bug,"[8] or it might have been connected, if co-opted by biologists, to a more specific class of animals. As it happens, events took the latter course, and now "insect" refers to a member of the class *Insecta* even when used by biological naïfs. (Everyone learns in elementary school that insects have six legs.) The reference of "insect" changed, then, from something less determinate to something more determinate,[9] consequently becoming more susceptible to analysis.

10.5 Conceptual Diversity

Different social groups—East Asians and Westerners, men and women, the young and the old, the extroverted and the introverted—occasionally have different concepts corresponding to the same words. Or that is one interpretation of the work in experimental philosophy demonstrating differences in case judgments among various demographic groups (Machery 2017, Table 2.9). Let me explore the implications for philosophical analysis.

One concern is that, under such circumstances, analysis can be only a parochial affair, the investigation of local folkways rather than eternal verities. That worry will be addressed in Chapters 12, 13, and 14. The question I want to pose here is whether conceptual diversity might stall analysis in the short term by creating irreconcilable differences in case judgments. When philosophers from diverse demographics work together or simply argue with one another about essential natures, will they find themselves stymied because what is in their variously constituted heads moves them to say conflicting things about key cases?

---

[8] There is a narrow biological term "bug" referring to the order *Hemiptera*, but I think that the ordinary English term is distinct; this is a case of polysemy.

[9] It may, as is often the case for terms for higher-order taxa, fall somewhat short of complete determinacy: there is arguably no fact of the matter whether the ordinary English term "insect" refers to the class *Insecta* or to the subphylum *Hexapoda*, despite the morphological magnetism of the former option. Even the technical Linnaean terms have, as every systematist knows, motile boundaries, especially as far as the classification of extinct organisms is concerned.

The pièce de résistance of experimental philosophy's diversity studies is the difference between what Westerners and East Asians say about the Kripkean Gödel cases. Westerners are more likely to have "causal-historical" intuitions; East Asians to have "descriptivist" intuitions. As I mentioned when reporting these studies in Section 4.4, however, the differences within groups are if anything greater than the differences between them. Approximately 60 percent of Westerners and approximately 40 percent of East Asians make "causal-historical" judgments—which means that both groups are split nearly down the middle. To invoke conceptual diversity in order to account for the breakdown of philosophical analysis, then, a multicultural perspective hardly seems necessary; the lack of progress might well be explained by looking solely at individual differences within a fairly homogeneous group, such as English-speaking pale male philosophy professors in their golden years.

No doubt putative differences of this sort help to account for certain philosophical impasses. But they cannot be the whole story: many of the most intractable obstacles to identifying essential natures present themselves even to an individual mind going solo. Professor Crusoe, cogitating in his hand-hewn armchair thousands of miles from the nearest department of philosophy, will run into the same problems with coffee-flavored dishwater, swan ancestors, and swan duplicates as the rest of us, which suggests that the principal difficulty in finding an analysis of water or swans is not a matter of interpersonal disagreement. The problem, apparently, lies within.

10.6 Conceptual Poverty

The case judgments critical to the enterprise of philosophical analysis, certain experimental philosophers have argued, are unreliable through and through. The culprit is our concepts, which simply don't have the resources to guide us to the truth when it matters most.

Machery (2017, Chapter 4) gives a straightforward argument for unreliability based on the sort of numbers reported in the previous section for judgments about Kripke's Gödel scenario. Even within a homogeneous group of Western subjects, just 60 percent judge that "Gödel" refers to

Gödel; the other 40 percent say that the proper noun refers to Schmidt, the true discoverer of the incompleteness theorem. Whatever the correct answer, at best 60 percent of those judging get it right. Those are not very good odds. Machery suggests that we are not much better off with a range of other famous thought experiments.

Another argument hinges on the ordering and framing effects described in Chapter 4. To refresh your memory of these effects, consider the original "Mr. Truetemp," an individual who has unerringly correct beliefs about the temperature without knowing why. When asked to judge whether Truetemp knows the temperature, people's answers depend to a certain degree on whatever question precedes the Truetemp question. If they have just been asked about a clear case of accidentally true belief, they are more likely to say that Truetemp knows the temperature than if they have just been asked about a clear case of knowledge. These data reveal that irrelevant factors exert a significant effect on a class of case judgments that has played an important role in the analysis of knowledge. Upon finding that our judgments are subject to such perturbations, experimental philosophers argue, we should cease to rely upon them (Swain et al. 2008; Machery et al. 2018). The same sort of argument can be driven by the variations in judgment among demographic groups discussed in the previous section, if those variations are interpreted as arising not from differences in the relevant concept but from irrelevant differences in cultural context.

Against these attempts to eat away at the foundation of philosophical analysis, it might be argued that we professional philosophers seem to get by pretty well with our case judgments, including many of those targeted explicitly by experimental philosophy. The masses may disagree 60/40 on Gödel cases, but the elect do not: they grasp that "Gödel" refers to Gödel and they have learned the apposite lessons about the nature of reference. Further, with due care and attention, they can overcome or at least minimize the effects of ordering and framing. Some gray cases may continue to generate controversy—fake barns, Chinese rooms, and so on—but because we have plenty to go on, we can make progress without having to rely too heavily on such disputable data. This is the "expertise" defense of analysis, according to which the effects uncovered by experimental philosophy, although real, have nugatory impact on the properly trained, well-prepared mind (Wright 2010; Williamson 2011; Rini 2015).

Experimental philosophers of the skeptical school do not deny that the profession has great confidence in its judgments and little expectation of disaster. But they see this as a symptom not of expertise but of complacency nurtured by institutional censorship of heterodox views about cases (Turri 2016).[10]

The best and also the most interesting way to adjudicate this dispute, I believe, is to uncover the psychological causes of variability and instability in philosophical case judgments. Such an inquiry, rooted in the nature of philosophical concepts, can perhaps answer the all-important question: have professional philosophers successfully cultured a sensibility that is able to resist the factors that disturb the judgments of ordinary folk, or have they rather created an academic monoculture that gives the impression of consistency and stability where there is little or none?

Edouard Machery's explanation of the diversity of case judgments suggests the latter answer. Most case judgments are stable and reliable, Machery is happy to concede (or else what use would our concepts be?). Further, philosophical analysts successfully use these reliable judgments to eliminate many prima facie plausible contenders as theories of essential natures. The rejection of the true belief theory of knowledge, the divine command theory of logical truth, the simple counterfactual and regularity accounts of singular causation (my examples rather than Machery's) are all perhaps warranted, and constitute genuine philosophical progress.

After this first theoretical culling, analysis becomes more challenging (Machery's skeptical story continues). The surviving theories agree on the ordinary and straightforward cases; where they diverge is with respect to the judgments they endorse about scenarios that are complex or unusual or artificial or tricky in various ways (Twin Earth, preempted double prevention, Kripke's Gödel, the Chinese room). To test the survivors against one another, then, it is necessary to have clear, stable judgments about such cases. It is here that Machery thinks our minds fall short, delivering uncertainty and inconstancy.

One reason for this is the difficulty of thinking about artificial cases; even professional philosophers, Machery suggests, may fail to fully understand the ramifications of "Jackson's Mary-the-neuroscientist case, . . .

---

[10] Machery (2017, 129), Table 4.1, shows how ordinary people disagree with philosophers' consensus case judgments about some famous scenarios.

10.6. CONCEPTUAL POVERTY

Locke's prince-and-the-cobbler case, brain swapping cases, Routley's last-man thought experiment, and Chalmers's zombies" (Machery 2017, 206).

The other reason, which I will focus on here, is that the complex test cases needed to discriminate among advanced philosophical theories "pull apart what usually goes together," in order to figure out which of the things that usually go together in fact determine category membership. This kind of testing of course makes perfectly good sense; it is the philosopher's version of the scientist's crucial experiment. But philosophy as opposed to science runs aground because our minds, and in particular our concepts of the relevant categories, do not contain enough information about the true criterion for membership to generate a clear judgment.

Machery and Seppälä (2011) propose a model of concepts to undergird this claim. In their view, any particular mental term—a concept in the "token structure" sense—is attached to multiple cognitive structures. These may be prototypes, exemplars, theories, or various other things.[11] Each provides its own criterion for category membership. But they compete rather than cooperating: if a specimen fits some criteria and not others, they do not work together to make the best call, but rather simply state their own opinion and then withdraw. The mind must do the best it can with these multiple judgments. In ordinary cases, the judgments will tend to agree, and so the multiple conceptual voices will sing in harmony. In the cases that look to decide philosophical disputes, however, they will give conflicting answers. There is, according to Machery and Seppälä, no "master criterion" to adjudicate among these answers. So the mind, if it is to deliver a determinate judgment, must choose a side without having any principled way of doing so. Often, says Machery (2017), the "superficial content" of scenarios will make a difference to the answer chosen. Disorder and instability are the consequences—and worse, unreliability, since the decisions that distinguish between philosophical theories are made in what are effectively arbitrary ways.

A variant of Machery's explanation is Paxton and Greene's (2010) suggestion that two distinct systems influence judgments about moral dilemmas such as trolley problems. One system is emotional while the other

---

[11] Machery and Seppälä think it important to maintain that the separate cognitive structures are distinct concepts, rather than parts of a single concept. I have given this claim my own interpretation in terms of token and cognitive structure.

is intellectual; they therefore fail to communicate, engaging in a mental tug-of-war rather than delivering a coherent joint verdict. Again, the consequence is unstable intuitions tipped one way or another by extraneous factors.

Finally, Machery (2017) suggests that instability or unreliability in judgments of philosophically decisive cases may be a result not of dueling criteria but of a single inadequately decisive heuristic. Suppose, for example, that attached to a concept are multiple criteria, just as Machery and Seppälä (2011) suggest, but that there is a weighting scheme to decide among conflicting judgments, resulting in a single master criterion for category membership. The scheme is not particularly fine-grained, however; consequently, when a split decision is rather close—many criteria weigh in on both sides of the judgment—the result is a stalemate.

You might suppose that in a case like this, the mind's reaction would be to declare a verdict of "uncertain." That would of course create its own roadblock to analysis, as Johnston and Leslie (2012) note. Machery (2017), however, postulates that the mind will tend to make a definite judgment call that is apt to swing back and forth under the influence of the immediate context. In either view, the prospects for analysis are grim: whether judgment about decisive cases is uncertain, unstable, or unreliable, it will be impossible to make philosophical progress past a certain point.

I want to look at these sorts of explanations of the failure of analysis—explanations that turn one way or another on the poverty of conceptual content—from the perspective of inductivism about conceptual structure. After a few general comments, a new explanation of diversity and instability in case judgments will suggest itself.

An inductive concept—or more precisely, the cognitive structure of an inductive concept—is a collection of beliefs, typically but not necessarily theoretically integrated. The beliefs as a whole function as a decision procedure for category membership simply by the application of inductive (and deductive) logic. They may be diverse, then, but there is no important sense in which they constitute separate criteria for category membership. There is only one (normally fallible and defeasible) criterion for membership: the beliefs should collectively give grounds for believing that the specimen in question is a category member. They may do so by pointing to category membership as the best available explanation (and a good one) of the specimen's observable properties, as the core causal beliefs do in the canonical cases of basic natural kind classification. But there are no restrictions: any logically permissible form of reasoning may play a role in inferring category membership.

The inductive picture of the mind is not consistent, then, with Machery and Seppälä's (2011) conception of multiple, separate cognitive structures. But it is certainly one way to implement the "heuristic" idea that cognitive structure implements a decision procedure that is not an ultimate and incontrovertible criterion for category membership. Indeed, that idea lies at inductivism's very heart.

To what extent, then, does the heuristic nature of classification using an inductivist concept threaten the viability of analysis? The chemical and biological beliefs found in experts' natural kind concepts are sufficiently pregnant with implication that they license unequivocal verdicts about an enormous range of cases, including many that might strike a beginner as problematic: there is little doubt among the well informed that penguins are birds or that mercury is a kind of metal. At the same time, expert category judgments are not unwavering in every case; the transitional forms that pave the way from swan ancestors to swans themselves are a case in point. Here even a fully developed swan theory may say, "I can't make a definite judgment."

In spite of this residual indeterminacy, inductivism gently undermines Machery's most pessimistic conclusions about the analysis of philosophical and other categories, for two reasons. First, as penguins and mercury show, there is nothing about the nature of inductive theories that systematically precludes their making confident classifications when specimens are unusual or pull apart things that usually go together. Machery's strongest argument is therefore held in check.

Second, inductive logic is an accurate judge of its limits. When the evidence is inconclusive, a clear-sighted inductive reasoner knows it. If they take a stab, they are aware that they are guessing. Thus you might expect, as Wright (2010) argues, that where a reasoner's concept is not up to the job of making an accurate classification, the reasoner will refrain from making a definite judgment rather than blundering blindly around the gray area while pretending that they know exactly where they are going.

The explanations of variability, instability, and unreliability in case judgments I have so far considered hinge on the diversity and poverty of the relevant concepts. But there is another kind of explanation that must also be considered: that these features of the judgments are caused by diversity and instability in logic itself.

10.7 Inductive Plasticity

Say that an inductive logic—under which heading I include any system of inductive inference—is *strict* if it allows no inferential latitude: two reasoners having the same evidence and using the same strict inductive logic will, if they reason correctly, reach more or less the same conclusions (or will attach more or less equal probabilities to the same hypotheses).

Philosophers of inductive inference tend to doubt that there is a uniquely rational, strict inductive logic.[12] Either we have a choice of strict logics, any one of which could reasonably be adopted, or if there is no choice, the uniquely rational inductive logic allows multiple inductive endpoints. In short, inductive logic is *permissive*.

In Carnap's inductive logic, for example, you may freely choose a value of the parameter λ, which determines the speed at which you learn from the evidence (Carnap 1950). In Bayesian epistemology, you may freely choose your prior probabilities, which determine pretty much all of your inductive behavior (Howson 2001). You might say that both Carnap's system and the Bayesian system are *inductive frameworks* that allow a range of strict *inductive logics*, any of which may be rationally applied in the course of ampliative reasoning (Strevens 2004).

Most categorization is inductive inference, says the conceptual inductivist: when we decide whether tears are a kind of water or whether a specimen of *Cygnus atratus* is a kind of swan, we reason inductively using our mental theories of waterhood and swanhood. If different people adopt different inductive logics, even within the same inductive framework, they may differ in their inductive conclusions and so in their classifications.

---

[12] White (2005) is not so sure.

Consider, for example, the heterogeneous categorical inclinations that Turri et al. (2015, 384–385) report for the following Gettier-like case:

> Emma purchases a diamond from a jewelry store and puts it in her pocket. A skilled jewel thief tries to steal it from her pocket before she leaves the store, and he succeeds. Someone secretly slips a diamond into Emma's pocket before she leaves the store.

Asked whether Emma knows that there is a diamond in her pocket, 55 percent of respondents responded affirmatively, attributing knowledge to Emma; 45 percent, by contrast, withheld the knowledge attribution.

Why the divergence in judgments? Could it be explained by a difference in inductive style? Perhaps the knowledge deniers are more cautious learners—perhaps they have a higher value for something like Carnap's λ—and so are more reluctant to infer the presence of knowledge from the usual diagnostic properties, justification and truth, given the absence of another diagnostic property, the counterfactual reliability of the putative knower's belief formation process, whereas the knowledge attributers forge ahead. Neither group is in general more reliable in its reasoning than the other; they merely exemplify different degrees of inductive risk-taking.[13]

Or consider the matter of molecular duplicates. Is a swan hatched from a swamp-spawned particle-perfect replica of a swan egg also a swan? There are strong reasons to decide both for and against membership in the class of swans. On the pro-membership side, the duplicate satisfies all the relevant core causal beliefs. It has (say) white feathers, a red beak, and the ability to swim, all for the same reasons, physiologically speaking, as real swans. Further, these physiological mechanisms—the whiteness-causing mechanism and so on—developed from the embryo in the same way in the duplicate as in a normal swan, and the replica breeds successfully with real swans. The usual causal minimalist argument schema seems to apply: the best explanation of these properties would be the specimen's swanhood, so you should infer that it is a swan.

---

[13] Colaço et al. (2014, 207) make a related suggestion to explain age effects in judgments about fake-barn cases; their account turns, however, on a controversial hypothesis about knowledge claims: "If knowledge attribution is stake-sensitive, then older people [thought to be more risk-averse] should be less likely to attribute knowledge in fake-barn cases."

But on the anti-membership side: the organism has the relevant features and mechanisms because of a freak chemical incident in the wetlands, not for the reasons that real swans have them—the latter having to do with the mechanisms of swan reproduction. This line of thinking does not challenge the factual claims of the pro-membership argument in the previous paragraph. It rather trades on a certain latitude in the entanglement relation. Swanhood is entangled with the whiteness-causing mechanism, according to our mental theory of swans. That implies that all genuine swans possess the mechanism for the same reason (as I will explain in Section 12.5). What reason might that be? If it is something to do with biological reproduction, then the anti-membership argument has the upper hand. But if it is merely something to do with development, then the pro-membership argument holds its own, since the process of development is identical in a naturally produced swan egg and in a molecular replica thereof.

The right answer in a dilemma such as this depends on how you implement inference to the best explanation. If we have variously chosen inductive parameters or logics that split on the implementation, we will disagree on the duplicate's swanhood (while recognizing that it is not the clearest of cases). And such disagreement looks to prevent our completing the analysis of swanhood.

I considered, in the previous section, the possibility that such impasses arise from conceptual poverty—that there is not enough content in our theory of swanhood, say, to speak to the question whether or not a swan's molecular duplicate is itself a real swan. The inductive explanation in the present section blames, by contrast, the looseness of inductive logic. Which story offers the better account of case judgments' diversity and instability?

My inclination is to think that plasticity has the more important role to play, because it does a better job of explaining why we positively disagree about some cases. I remarked in the previous section that conceptual poverty, at least of the inductivist variety, should leave us uncertain rather than opinionated. Our theories fall short of making a definite judgment, and we know it. We and our philosophical opponents ought therefore to agree that certain thought experiments are not going to elicit intuitions powerful enough to settle our dispute.

Truces of this sort are not unknown. More characteristic of philosophical hurly-burly, however, is the "clash of intuitions," in which there is disagreement, sometimes rather sharp, as to the right thing to say. Plasticity, by allowing two reasoners to reach contrary conclusions from the same premises, can account for these differences of opinion.

At the same time, something like plasticity can explain transient effects such as variations in judgment due to ordering and framing: differing proximal contexts make different factors more salient, leading (so I hypothesize) to their being weighted more heavily in the ensuing inductive deliberation. If the rearranged weights are among those permitted by the canons of inductive logic, you could understand what occurs as an ephemeral and largely unobjectionable transition from one permissible inductive scheme to another. Even short-term instability in categorization might, then, be understood as a harmless side effect of the plasticity of inductive reasoning.

10.8 Do Poverty and Plasticity Explain the Failures of Analysis?

Complexity, diversity, poverty, plasticity: these are the obstacles to philosophical analysis examined in this chapter. In the light of inductivism, complexity was sidelined; diversity, meanwhile, is something to save for the discussion of the substantiality of philosophical knowledge in Chapters 12, 13, and 14. It is poverty and plasticity, then, that are given all of my attention here.

I have brought conceptual inductivism into play, but not yet the dispositional approach to reference. How, then, does dispositionalism modulate the poverty and plasticity approaches to explaining the failure of analysis?

Begin with poverty. As I remarked in Section 10.6, the inductivist's version of the poverty argument posits that the beliefs making up many philosophical concepts are insufficiently rich, broad, or precise to deliver determinate judgments about a wide range of dialectically important "hard cases." Analysts are left mired in uncertainty.

If the dispositional approach to reference is correct, however, then there are no hard cases. When the question of membership cannot in principle be decided, it is simply indeterminate whether the specimen in question is a category member or not (which is to say that the specimen is determinately a borderline case). The situation that Machery fears—that a cognitive shortfall leads to a failure to discern facts about category membership essential to deciding between rival theories—cannot, therefore, arise. Two competing philosophical theories, one of which adamantly predicts that the case is a category member and one of which predicts that it is not, are both wrong. The correct theory will predict the specimen's indeterminate or borderline status. Analysts thus have more or less all the information about the case relevant to analysis. They are not impoverished at all.[14]

Next, plasticity. While inductivism predicts that conceptual poverty will express itself in uncertain judgments, it predicts that inductive plasticity will express itself in diverse judgments, with different philosophers drawing different conclusions about the same case. There is, in short, disagreement.

If the dispositional approach to reference is correct, it might appear that such disagreement is all a regrettable mistake. Assuming that our differing inductive dispositions are stable, our categories have slightly different extensions. I am right about the membership of my category and you are right about the membership of yours. As long as we argue about case judgments, we are talking past one another; if we were enlightened as to the nature of reference, we would abandon our argument in order to pursue our own personally valid philosophical analyses.

That is not a particularly attractive picture. It is also, I believe, misconceived, for reasons that emerged in the discussion of the elm/beech scenario in Section 9.6. You and I both aim to be good linguistic citizens, meaning that we intend to use, or assume we are using, our words in the same way as other speakers of English—including each other. As a consequence, a situation in which, even as a total set of evidence comes in, we continue to apply our words to different things, is potentially unstable. If we adhere to dispositionalism, it is downright incompatible with the good citizen assumption.

The instability could work itself out in one of two ways. Either we agree to regard the troublesome specimens as borderline category members, or one judgment somehow comes to dominate collective usage. Whichever

---

[14] An analyst who does not endorse dispositionalism might worry about conceptual poverty—they might worry that some of their indeterminate judgments fail to disclose important facts about category membership—but they are wrong.

path is taken, two problems are solved. First, there is not some ineffable truth about the troublesome specimen that evades us, and in doing so prevents us from finding the essential natures that we seek. Second, our ultimate dispositions to categorize are not misaligned, and so our earlier disagreements were not mere miscommunications.

To sum up, once dispositionalism is taken into account, it is far from clear that either conceptual poverty or inductive plasticity could deprive philosophical analysts of crucial information about category membership. That is one reason to believe that poverty and plasticity have not severely handicapped analysis—and so that its troubles are to be otherwise explained.

Here is another such reason, picking up on the notion of philosophical expertise described in Section 10.6. It has often been remarked that professional philosophers' judgments may diverge from beginners' judgments about category membership. Professionals tend to agree that Putnam's X_yZ is not water (Section 9.6). Beginners frequently think otherwise; some time in the classroom or tutorial chamber, however, can induce them to change their minds. The same may be true of Gödel cases and Gettier cases. And of swans, too: novices are more likely and professionals less likely to count a molecular duplicate as a swan (though here experts' knowledge of scientific systems of taxonomy that explicitly require an evolutionary connection may confound the issue).

In the orthodox story, the novice to expert shift represents progress: experts are simply better at thinking through their case judgments than novices. That story might seem rather suspicious, in the light of many theories of concepts and case judgments. If case judgments are simply a matter of "intuition," or assessing the fit between a specimen and a prototype, why should a PhD in philosophy render them any more accurate? Philosophical experts are not like scientific experts: they do not have more information than novices about their proprietary categories. Setting out on their analyses, they know as much and no more about knowledge, causality, and reference as the next person. Thus, as I reported in my earlier discussion of expertise, Turri (2016) and many other experimental philosophers have hinted that professionals' "correct" judgments result from indoctrination and exclusion rather than any relevant proficiency.

Inductivism provides an effective comeback to the defender of philosophical expertise. Categorization, especially for difficult or complex cases, is inductive reasoning. Inductive reasoning is hard. Experts in argumentation will do it better. In this respect, the shift from novice to expert judgments about X_yZ or duplicates is like the shift from novice to expert judgments about raccoons that have been dressed up as skunks: in Keil's experiments, as I explained in Section 5.3, what the 5-year-olds who count the transformed animals as skunks lack is not the adult concepts of raccoon and skunk, but the ability to think in a sophisticated way about the inductive consequences of the concepts—that is, the theories—that they already possess.

If this line of thinking is correct, then perhaps the experimental philosophers are exaggerating the extent to which we must throw up our hands when we encounter 60/40 splits in ordinary people's judgments about Gettier or Gödel scenarios. The professional consensus in these cases may well (as a number of proponents of the "expertise defense" cited in Section 10.6 have argued) show the way to the right answer. Such a conclusion would certainly help to relieve my own confusion at feeling "case certainty," or something close, about some of these scenarios in spite of the wild variability of regular people's responses.

A third reason to think that poverty and plasticity are not the prime culprits in the derailing of analysis is that their putative explanation of analysis's troubles does not quite correspond to the facts. Arguments or uncertainty about case judgments are not the principal difficulty in analyzing the water category, or for that matter in analyzing knowledge or singular causation. The problem is not so much a lack of solid case judgments as too many: it has simply proven fiendishly difficult to find hypotheses about essential natures capable of accommodating all of these facts about what falls into what category.

A quite different approach is therefore required to explain the fate of analysis—an approach that focuses not on a supposed shortage of the intuitions needed to zero in on essential natures, but on the possibility that there are no essential natures on which to converge.

CHAPTER 11

Against Essential Natures

Suppose that, as referential dispositionalism implies, your best theory of swans is a perfect categorizer of swans—it makes the right decision, in every instance actual and possible, as to whether a specimen is a swan, is not a swan, or is a borderline case. There are several metaphysical theses that attempt to spell out the essential nature of swanhood and that are compatible with these judgments; which of the theses is correct? If swanhood has no essential nature, there is no answer to this question.

Make the same assumption about water, that your best theory categorizes any specimen correctly as either water, non-water, or as a borderline case. Any extensionally viable thesis about the essential nature of water—any thesis that reproduces the correct judgments about what is and is not water—is peculiarly complex and makes reference to the kitchen hand's plans and other prima facie unsuitable determinants of waterhood. Does that mean that the essential nature of waterhood is complex and makes reference to psychological facts? That whether something is water depends on its satisfying some complex, partly intentional formulation? Maybe not; maybe it's rather that water has no essential nature.

How could that be? How could categories such as swanhood and water, categories with (largely) determinate extensions and considerable practical and explanatory significance, turn out to have no essential natures? This chapter explores three possible explanations.

11.1 Regress Arguments and Primitive Concepts

In attempting to account for the failures of the philosophical analysis of knowledge, Timothy Williamson writes:

> One would not expect the concept *knows* to have a non-trivial analysis in somehow more basic terms. Not all concepts have such analyses, on pain of infinite regress; the history of analytic philosophy suggests that those of most philosophical interest do not. . . . Attempts to analyze the concepts *means* and *causes*, for example, have been no more successful than attempts to analyze the concept *knows*, succumbing to the same pattern of counterexamples and epicycles. (Williamson 2000, 31)

Rather than attempting to pinpoint exactly what Williamson has in mind, let me present two distinct lines of thought that might be developed from this passage.

The first focuses on essential natures. A complex essential nature must be composed of properties with metaphysically more basic natures—on pain, as Williamson writes, of infinite regress. As the modern analysts noted, some properties may be interdefined, so that they appear in one another's essential natures, but if they are to have any foothold in reality, the clusters of definitions must also be connected to more basic properties, and ultimately to the most basic properties of all.

Could knowledgehood be one of those most basic properties? On the face of things, it seems unlikely: the metaphysically basic properties will surely be the fundamental physical properties (perhaps along with the fundamental phenomenological properties, if the dualists have it right). At the very least, I would need a rather creative story to persuade me that knowledge could be among the basic properties.

The second line of thought goes by way of concepts rather than moving directly to essential natures. Its first premise is the compelling idea that, in order to acquire new concepts, we must in some sense build them from preexisting concepts. The second premise is what you might call the conceptual analyst's thesis, according to which the philosophical analysis of a concept exhibits the blueprint of this building process, showing how the new concept was constructed from the preexisting concepts. It follows from these ideas that any concept not built from other concepts—a conceptual primitive, one of the units from which all other concepts are built—has no analysis. If philosophical concepts such as knowledge and causality are among the conceptual primitives, then, they have no analyses and therefore no essential natures. (Or at any rate, no nontrivial analyses and therefore no nonprimitive essential natures.)

It is far easier to accept off the bat that philosophical concepts are conceptual primitives than that the corresponding categories or properties are metaphysical primitives. Nevertheless, by this argument the conceptual premise is transformed into the metaphysical conclusion.

As you will anticipate, I reject the conceptual analyst's thesis. If philosophical analysis is inductive analysis, then there is no straightforward relation between a concept's cognitive structure and its essential nature. The cognitive structure is a set of theories and beliefs that contain information about the corresponding category. These representations constitute clues to the category's essential nature—the only clues we have—but (except in unusual circumstances) they do not specify the nature.

Inductivism harmonizes happily with the premise about conceptual construction: concepts are acquired by building theories, and theories are built out of preexisting representations, hence preexisting concepts. (An important subtlety will be explored in Section 15.1.) But what is built is not a representation of the corresponding category's essential nature; it is rather a set of ordinary beliefs about the category, beliefs that make no special semantic or metaphysical claims. There is therefore no philosophically interesting sense in which the preexisting concepts are, in virtue of their role in conceptual construction, more basic or primitive than the constructed concept. They are simply prior or older—and (I would hazard) contingently so at that.[1]

I remarked in Section 10.4 that a concept with a highly complex cognitive structure might have a rather simple essential nature. Likewise, a concept that is not built up from other concepts—an ultimately prior or innate

---

[1] A preexisting concept might even turn out, on analysis, to be less basic than a concept it helps to construct. The constructed concept, that is, might figure in the correct analysis of the preexisting concept.

concept—might have a rather complex essential nature. Even if the philosophical concepts are unbuilt builders, then, they may have interesting, informative analyses. Considerations of regress in conceptual construction therefore cannot account for the intractability of philosophical analysis.

11.2 Idemic Essential Natures

For every category, there is a categorical property: for the swans, swanhood; for water, waterhood; for the prime numbers, primeness; for the causal, causality. In some cases, the categorical property can be identified with some other property. Primeness, for example, can be identified with the property of a number's being divisible only by 1 and itself. Gas pressure can be identified with the force per unit area exerted by the collisions of gas molecules on a surface.

When such an identification is possible, the property with which the categorical property is identified is a natural endpoint for analysis; it is the category's essential nature in what I will call the *idemic* sense (from the Latin *idem*, meaning "same").[2] To state a category's idemic nature is plausibly not merely *a* goal but *the* goal of analysis—the gold standard for the analytic enterprise. In this section, I use the examples of water and swan to explore some reasons that categories may lack idemic natures, thwarting the analyst's fondest hopes.

This is an easy job, you might think, for the inductivist. Idemic essential natures are fixed by conceptual definitions or perhaps intensions; if no such things exist then there are no idemic natures. But as the very notion of inductive analysis supposes, it is possible to reach conclusions about essential natures that are not simply read off the contents of the mind. It is possible, that is, to find an identity in the world that is not prefigured by a definition in the mind or by some other conceptual structure.

Philosophers of science call such discoveries "theoretical identifications." Gas pressure is a case in point. Scientists are quite happy to say that gas pressure just is the force per unit area induced by molecular pounding.

---

[2] This notion of an idemic essential nature perhaps ought to be strengthened: there may be a reductive element to the quest for essential natures that requires that an idemic essential nature be a complex of properties that are in some sense more basic than the categorical property. I impose no such requirement here. There is more about reducibility in Section 15.4.

Clearly, however, our concept of gas pressure is not, or at least before the formulation of the kinetic theory of gases was not, built around such a definition. The same might be said of our concepts of weight and electric current. When we came to identify these properties with, respectively, mass times gravitational acceleration and flow of electrical charge, we were learning something from empirical evidence rather than reading from a definition preinstalled in the mind, but they are genuine identities all the same.[3]

How do theoretical identifications involving concepts that are inductive, and thus have no definitions or stipulative intensions, come about? We endorse identities, I suggest, when there turns out to be a single property that plays the causal role specified by our category-involving hypotheses (and provided, of course, that we are sufficiently confident that those hypotheses are true). Molecular pounding plays precisely the role we accord to pressure, so we take pressure to be nothing over and above molecular pounding.

Such an identification is not a matter of logical deduction. But we make it all the same—because, I think, we are committed to inference to the best explanation, or some similar rule of inductive inference, and we admit identifications as explainers. Why does gas pressure increase with temperature? Because the average speed of particles bombarding container walls increases with temperature, and gas pressure just is the force exerted in the course of that bombardment. The identity posit is essential to the explanation; inference to the best explanation, then, from time to time leads us to endorse such posits.[4]

This suggests a general account of the circumstances under which categories have idemic essential natures: a category has essential nature P just in case P plays the explanatory role that we attribute, upon learning a total set of evidence, to the categorical property. Waterhood has a nonprimitive idemic nature, for example, just in case there is some compound property that has the explanatory role that our completed theory of water ascribes to waterhood.[5]

---

[3] These ideas are familiar to many philosophers from Kripke (1980), although the notion of a theoretical identity is much older.

[4] On the matter of identifications as explainers, I note in passing that in other work, I have proposed that the causal explanation of empirical regularities operates by positing metaphysical identities (Strevens 2008a, §7.6).

[5] The relevant explanatory role need not be a causal role; it might be any "primary" explanatory role, in the sense to be characterized in Section 13.1.

This proposal is, however, too narrow. Consider a concept that is founded on a definition. The idemic essential nature of the category must surely be whatever property is picked out by the definition. We may not, however, ascribe the category any particular explanatory role; indeed, we may not think of it as participating in the explanatory economy at all. So idemic natures are not always determined explanatorily. Further, it is at least conceivable that two different lower-level properties play exactly the same explanatory role as a given high-level categorical property. But a category cannot have two distinct idemic natures; a theory of idemic natures must, then, lay down a principle for adjudicating such cases. (No essential nature? A "disjunctive" nature?)

In order to deal with these concerns, I will propose a dispositional approach to idemic natures, modeled on the dispositional approach to reference. A category has an idemic essential nature P, on the dispositional approach, if upon acquiring a total set of evidence—a set of observable facts sufficiently complete that no further facts would change our beliefs about the category—we would, given enough time and care, identify the categorical property with P.[6] You might understand the dispositional approach as attempting to spell out what it is to be an idemic nature, or you might understand it as a placeholder, to be replaced eventually by a set of non-dispositional criteria for being an idemic nature that imply the dispositional connection, much as various theories of reference imply the dispositional doctrine of reference. The argument that follows will go through either way.

Now to pose the key question: might basic natural kinds such as water and swan lack idemic essential natures?

Let me proceed using the example of water. To qualify as an idemic nature, I assume, a property complex must at a minimum be extensionally equivalent to the categorical property, which is to say, the complex must be instantiated by all and only the category members.[7] As we have seen in the case of water, it is very difficult to devise a property complex that

---

[6] Were some thinkers to identify the categorical property as P and others to identify it as a different property Q, it would follow that these two groups were all along thinking about two distinct, though perhaps extensionally equivalent, categories.

[7] Intensional equivalence is also presumably required, but I will not lean heavily on that requirement here.

satisfies this extensional criterion. Might the criterion be impossible to satisfy? No; there will always be at least one extensionally adequate complex, the metaphysical mirror image of our last, best theories of water. Let me explain.

For simplicity's sake, suppose that there is a unique ultimate theory of water on which all incipient minimalist theories of water are bound to converge upon receipt of a total set of evidence. The extension of the concept of water is determined by the form of this final theory together with the dispositional doctrine of reference: a substance is water just in case an ideal inductive reasoner armed with the theory classifies it as water. (Ignore, also for simplicity's sake, the complications introduced by the plasticity of inductive logic.) Borderline cases of waterhood are determined in much the same way (Section 9.4).

The ultimate theory of water and the principles of inductive logic together determine a decision procedure, then, that makes unerring water classifications: presented with any specimen, the procedure will infallibly judge whether it is water, non-water, or a borderline case. The procedure will be propelled by inductive logic, and so any description of the procedure will be rife with references to evidential relations, subjective probabilities, or similar notions. But it should be possible to remove all such references to inductive reasoning from the criterion while preserving its decisions. Various inductive quantities (level of confidence, degree of evidential support) will thereby be transformed into a (typically nonlinear) metaphysical weighting scheme. The result of this process of "epistemic denaturing" is a criterion, perhaps very elaborate, that correctly tells you, on the basis of a substance's properties and surroundings, whether the substance is or is not water.

Such a criterion can be used to determine a property complex—in the case of water, a perhaps fantastically intricate property complex—that is necessary and sufficient for waterhood, namely, the complex the instantiation of which is necessary and sufficient for a substance to be classified by the criterion as water. Call this the *water complex*.

The water complex obviously passes the extensional test for an idemic essential nature for water. (It was constructed precisely for this purpose.) Can waterhood, then, be identified with the water complex?

The answer is affirmative, according to the dispositional approach to idemic natures, just in case we would make that identification if a total evidence set were to arrive. In what follows, I argue that we would not. The best candidate for water's idemic essential nature is not good enough to qualify as such. Thus, water has no essential nature.

To make things easier I will assume that our present theory of water is close enough to the ultimate theory as to make no difference between the two, either in their classifications of waterhood or in their power to incline us to make theoretical identifications.

Recall from Section 10.2 some features of our decision procedure for waterhood. First, the procedure is complex and draws on many diverse facts about the specimen to be categorized. Second, some of these facts concern extrinsic matters that seem not to have much to do with chemistry, such as the kitchen hand's intentions when imparting a distinct coffee taste to a large quantity of hot H_2O. Third, the procedure does not yield simple yes or no judgments. At the very least (and for the sake of the argument, let me suppose also at the very most) it attaches epistemic probabilities of some sort to its decisions. Many specimens will be rated as water or non-water with near certainty, but some will arrive with intermediate conviction: "probably water," "probably not water." (I will ignore borderline cases, which elicit a "not sure.")

All three features motivate our reluctance to identify waterhood with the water complex. Its tortuous structure is the least important, I think. We may well expect our essences to be elementary, but the disappointment of this presumption is hardly grounds for rejecting a hypothesis about essential natures when there is no prospect of a simpler alternative.

The kitchen hand's intentions are a different matter. We are pretty sure that waterhood is an intrinsic property of water, or that if it is relational, it is chemical relations alone that matter. The idemic nature of water, then, cannot have anything to do with what's running through the kitchen hand's head, so the water complex is a nonstarter as a candidate for water's nature.[8]

Let me sketch an explanation of these convictions. We are antecedently convinced that all of the chemical properties of water are intrinsic, or

---

[8] Some thinkers will, I suppose, be more friendly to a view in which our practical interests in dealing with water are built into the semantics of our water concept, or to a pluralistic view in which "water" picks out many categories, some practically and some chemically natured. But many will not, putting out of reach the level of consensus necessary for the water complex to count as an idemic nature.

11.2. IDEMIC ESSENTIAL NATURES

inhere in its chemical relations with other substances. This supposition plays an important role in inductive thought about water: it tells us that when looking for the mechanisms underlying the core causal beliefs, we can confine our attention to chemical mechanisms (Strevens 2007, §4). The idemic nature of waterhood is itself (should it exist) a chemical property, so it too must inhere in purely chemical facts—thus we reason. I call this an explanation sketch because it involves a tight circle of notions: chemical properties, chemical relations, and chemical mechanisms, which I will not try to unweave and show to be well founded (though I will have a little more to say in Section 14.5). What matters for my argument is in any case something that is difficult to deny, whatever its ultimate explanation: we, or most of us, are pretty sure that the essential nature of water (if any) is a purely chemical matter, and so we will reject, or at the very least be reluctant to accept, a metaphysical thesis that has a substance's waterhood constituted in part by the kitchen staff's psychic makeup.

The third and final impediment to identifying waterhood with the water complex is the unnaturalness of the epistemic denaturing by which the complex is determined. What goes into a decision procedure for waterhood? Absolutely anything that bears evidentially on the question whether something is water. That the kitchen hand did not intend to make coffee, for example, is evidence that a mix of H_2O and coffee solids is not coffee, and so it is evidence in favor of its being water. (This in the same way that the fact you ordered coffee is pretty good reason to believe that what is about to be brought to your table is coffee rather than some other drink.) The kitchen hand's intention may therefore quite reasonably feature in our deliberations when deciding how to classify the liquid in question.

Further, the intention will continue to be relevant even in a final theory of water, that is, a theory conditioned on a total set of evidence. This is due to the inductive nature of the water concept: because there is no proper part of the final theory of water that supplies conclusive categorizations—no definition or stipulation that trumps all other considerations—anything that has bearing on the question will retain some deliberative weight to the very end.

Epistemic denaturing converts all such evidential considerations into metaphysical constituents: anything that by the lights of the final theory provides reason, however slight, for or against believing something to be water is thereby apotheosized as a determinant of, indeed a part of, waterhood.

But this is rebarbative: any reflective person can see that, whatever the essential nature of a category, there will on occasion be reasons for and against classifying something as a category member that go beyond that nature. When epistemically denaturing a decision procedure for category membership, then, you can be almost certain that you are introducing matters into the putative idemic nature that are epistemically relevant but metaphysically immaterial. You can be almost certain, in other words, that you are generating a grotesque travesty of the category's real essential nature, if there is any such thing.

For some or all of the above reasons, a good inductive reasoner will reject the only extensionally adequate candidate for a theoretical identification with waterhood, concluding that water, although it has a determinate extension (give or take some haziness at the boundaries), has no idemic essential nature. By the dispositional characterization of idemic natures, it follows that water indeed lacks such a nature.

The attempt to determine the essential nature of swans will unfold in a related way, and with an equally bleak conclusion. On the one hand, a good inductive reasoner will refuse to identify swanhood with a property complex obtained through epistemic denaturing, for the same reasons as in the case of water.

On the other hand, there are, I have generously allowed, a number of other extensionally adequate criteria for swanhood, derived from the molecular, phylogenetic, and evolutionary approaches to biological taxonomy discussed in Section 10.3. Could considerations other than extensional adequacy single out one of these candidates as the true idemic nature of swanhood? Perhaps a certain candidate is inductively favored because it is markedly simpler than the rest? The dispositional approach allows such considerations to play a role in determining facts about idemic natures, but in the case of swanhood, that will not provide much help. Any of the major candidates is a plausible contender.

A choice could be made at random, but such a procedure is unlikely to give you the correct answer, if there is a correct answer. So a good inductive reasoner will decline to identify swanhood with any of the extensionally adequate candidates. From the dispositional approach, it follows immediately that swanhood has no idemic essential nature.

A final sally: Suppose that, in addition to having all the relevant biological evidence, I know the dispositional posit about idemic essential natures. Then I know that simply by choosing a particular identification from the pool of available candidates, I can make it so: my choice becomes the true essential nature of swanhood and, ipso facto, I have chosen correctly. That is the power of reflexivity.

Belief, however, is not voluntary. I would love to know the true nature of swanhood, and I know that by forming a belief about the matter I can fulfill my desire. Yet I cannot take that final doxastic step. The nature of swanhood is dictated, not by my will, but by the structure of the biological world, my beliefs, and my inductive dispositions—none of which I can directly control.

11.3 Explanatory Essential Natures

A category's essential nature was characterized in Chapter 1 as the property in virtue of which category members fall into the category—in the case of water, for example, whatever property specimens of water share, in virtue of which they are water. To put it another way, a category's essential nature is that property the possession of which explains category members' category membership.

These words suggest a conception of essential natures strictly weaker than the idemic notion. Even if we are reluctant to make theoretical identifications for all the reasons given above, there is surely all the same an explanation for why our categories contain the things they do—an explanation that can be read straight off the theory of reference. Call essences in this extension-explaining sense *explanatory essential natures*.

In some theories of reference, a category's explanatory nature is explicitly represented in the head. Traditional descriptivism, for example, posits a mental representation—say, "white-feathered, red-beaked aquatic bird"—that fixes the reference of the term "swan." It is a specimen's satisfying the description that explains why it falls into the extension of "swan"; the explanatory essential nature of swanhood, then, is the property picked out by the description: the property of being a white-feathered, red-beaked aquatic bird. Characteristically, perhaps inevitably, the property is also the category's idemic essential nature.

According to the complex of views advocated in this book—conceptual inductivism coupled with the dispositional approach to reference—there is nothing that plays this explicit, reference-fixing role. Yet—so it may seem—there must be explanatory natures, since reference can hardly be inexplicable. Let me develop this thought by assuming, for the sake of the argument, the stronger version of the dispositional approach to reference, namely, the dispositional theory according to which reference is determined by, rather than merely reflected in, ultimate categorical dispositions.

According to the dispositional theory of reference, for a substance to fall into the extension of the water concept is for a fully informed reasoner to count the substance as water. Such a decision can of course be explained, in which case the substance's falling into the category of water—its *being* water—can be explained. (If you recoil at the thought that category membership is explained by the theory of reference, then you will not regard "explanatory essential natures" as genuine essential natures. This part of the discussion is not for you.)

More specifically, any specimen that counts as water possesses some complex of properties in virtue of which the informed categorizer's theory inductively urges the conclusion that it is water. Now, take that property complex, for every specimen, and form a big disjunction (though not so big, if many specimens are categorized as water for much the same reason). The resulting disjunctive property—which is of course none other than the water complex introduced above—is shared by every specimen of water, and is the property in virtue of which it counts as water. The water complex is, if not water's idemic essential nature, then at least its explanatory essential nature.

The same strategy can be applied to the category of swans. There is some property complex in virtue of which any particular bird is categorized, by the fully informed theory of swans, as a swan. Take that property complex for each swan—in no case, perhaps, identical to any of the top contenders for the essential nature of swan according to a more traditional approach to species metaphysics—and form a disjunction. That is the essential nature of swanhood in the explanatory sense. Observe that the complex so obtained is unique: other property complexes that look more like traditional analyses might successfully capture the extension of swanhood, but they will not be the psychological explainers of fully informed categorization and so (according to the dispositional theory of reference) they will not explain category membership.

Why are we reluctant to count these complexes as essential natures? Why do we not recognize them as the analyses of water and swanhood that we were looking for? Perhaps we are, in our inductive inquiries, interested in idemic natures only. Then again, perhaps we are making a mistake: perhaps we are failing to recognize the essence-creating consequences of referential dispositionalism.

In the latter case, we may complete our philosophical analyses by taking the following route to enlightenment. We obtain full knowledge of the basic structure of the material universe—by inference from a total set of observable evidence—but we are frustrated in our pursuit of full philosophical knowledge, seeing that our theories of water, swan, and so on are merely inductive, neither specifying nor even insinuating the ultimate nature of waterhood and swanhood. Then we come to know the dispositional theory of reference. (The knowledge need not be incorrigible or certain.) Dispositionalism delivers a blow to the head, sparking the realization that the natures are already inherent in the theories: they are nothing more than the properties in virtue of which the theories make the classifications and so draw the categorical boundaries that they do. We see that there is no higher truth; there are just our everyday linguistic and cognitive tools.

The vision is a false one. As I have characterized a category's essential nature, it explains the category membership of every category member. But the disjunctive properties described above explain nothing. Any given category member has its membership explained by some part of the disjunction, but the explanatorily relevant disjunct varies from specimen to specimen and is never the same as the whole. There is no universal explainer of category membership, and therefore, there is no explanatory essential nature.

To better appreciate this point about explanation and disjunction, consider the many European wars that France has fought since 1750. The outbreak of each has a distinct explanation, even though the explanations to some extent overlap (debt, alliances, insecurity). Suppose someone were to say that there is a single explanation that accounts for every one of these wars; namely, the disjunction of the explanations of each. You would dismiss the claim out of hand. Each war started for its own reason; the disjunction of these reasons may be a useful summary but it is not an explanation in its own right. Grasping the disjunction in addition to the disjunct in no way improves our understanding of, say, the war of 1870, and grasping only the disjunction diminishes our understanding, because we do not know which of the disjuncts tells the explanatory story about the beginning of that particular war.

The same is true for the different specimens of water. The explanations why each is classified as water, detailing the courses of inductive reasoning that lead to the various judgments of category membership, share many elements. Within certain classes of specimens, the judgments are in every case explained identically; among different classes, however, there is only a partial overlap of explanations. To take the disjunction of these explanations is no more psychologically explanatory than it is historically explanatory to disjoin war stories.[9]

Summing up, a category such as water has neither an idemic nor an explanatory essential nature, roughly because even the ultimate theory of water—the theory held by a highly competent reasoner after swallowing a total set of observable evidence—makes its categorizations inductively and on highly heterogeneous grounds. That the ultimate theories of water and swans have this aspect is in large part explained by the fact that the concepts of water and swan are inductive. But not every inductive concept is hostile ground for essential natures. It is very likely that the concept of gas pressure is inductive, originally consisting of ordinary hypotheses about the causes and effects of pressure. In that case sufficient evidence led to a theoretical identification of pressure and molecular pounding, and thus an idemic essential nature for the inductively conceptualized category. There is more than inductiveness, then, to the story of how water and swan evade the analyst's net. I pursue the explanation further in Section 15.4.

Can the notion of essential nature be weakened further to jettison the requirement that natures explain category membership? Why not define an essential nature as the property that category members have in common,

---

[9] Any good theory of explanation must, then, individuate explanations. A theory of causal explanation, for example, must individuate causal mechanisms or something similar (Strevens 2008a, §3.7)—though as Franklin-Hall (2014, 2016) points out, not all philosophers of explanation have been careful to recognize this need.

or better (since the case of swans shows that there may be more than one such property), the disjunction described above—the disjunction of the category membership explainers for each specimen? You might call this the *extensional* notion of essential nature. Even the categories of water and swan have natures in the extensional sense, as does every category to which we can explicably refer—which is to say, presumably, all of them. Does that undermine the "no essential nature" account of the failure of analysis?

I think not. Water's and swanhood's extensional natures exist, but they are natures only in a trivial and insipid sense. They are not of sufficient philosophical interest to attract us as analyses; even were they to be spelled out for us, we would reject them as suitable targets of inquiry.

11.4 The End?

Suppose, then, that water, swan, and other basic natural kinds lack essential natures in any interesting, substantive sense. Suppose further—and this is now speculation—that the analysis of philosophical kinds has foundered for the same reasons. Knowledge, perhaps, is only too much like water: the sole property complex capable of capturing the category's extension is the one derived from the epistemic denaturing of the final theory of knowledge, yet that complex is disqualified as an idemic essential nature by its invocation of manifestly irrelevant factors and disqualified as an explanatory essential nature by its disjunctivity.

That would provide a compelling explanation of the disappointments of philosophical analysis to date: we have failed to find the essential natures of many philosophical categories because there are none to find.

From a pragmatic point of view, the loss of natures is no disaster. Our dispositions to categorize are, if not fully determinate, decisive enough to give natural and philosophical kinds extensions that serve our practical purposes. As to our theoretical purposes, the nonexistence of natures would have some salutary implications. We would no longer have to agonize over the many good theories of swanhood; we could say "There's no matter of fact" and go on to other things. Likewise, confronted with the convoluted and unnatural moves needed to construct an analysis of waterhood, we could get up and walk away from the game, declaring that there is nothing to know beyond the extension (or intension) of the term "water" and its corresponding concept.

And yet—the game we are quitting is philosophical analysis itself, the pursuit of substantive knowledge about the essential natures of things.

Is that it? After the promise of the previous chapters—analysis is inductive, and so tells us genuinely novel things about the philosophical categories; reference is reflexive, and so what we are told is real knowledge—is the route impassable, the journey over?

Philosophical analysis is perhaps doomed. Yet there is a reason that I opened this book with a chapter titled not "Philosophical Analysis" but "Philosophical Knowledge." We cannot discover the essential natures of either the basic natural kinds or the philosophical categories if they have no such natures, but we can discover other things that are valuable and substantial. We can discover that water's properties hinge on its being mostly made of H_2O, that swans share a certain ancestry and certain genetic signatures, and—so I will argue in the next three chapters—we can discover important things about the stuff of causality, rationality, and knowledge itself.

CHAPTER 12

Substance: Basic Natural Kinds

12.1 Reflexivity and Emptiness

Knowledge that is substantive and novel can come to the occupant of an armchair: the ruminating detective, putting together the clues gathered over the course of their investigations, may discover the identity of the murderer through the exercise of pure thought alone. That thought is acting, however, on substantive, novel information acquired the empirical way—interviewing the maid, eavesdropping on the butler, seducing the gamekeeper.

Armchair philosophy takes as its raw material, by contrast, not something that is new and interesting itself, but rather something—an initial or naive theory of the category in question—whose status as a foundation for inquiry is secured by fixing the facts about reference so that the beliefs that make up the theory are foreordained to be truth-directed. If reference is reflexive, then, analysts do not so much gather their clues from the environment as plant them there.

Or to put it another way, we philosophers are taking out of the world only what our minds earlier put into the world. We are less like the fictional detective than we are like the detective's creator, who seeds the earlier chapters with clues knowing all along how the story will end. Well, not quite—we don't really know how the story will end—but is the endpoint of philosophical inquiry any less contrived as a consequence? Is it not, all the same, the outward projection of our initial naive theory, rather than some weighty, independently interesting truth waiting out there in the world to be ascertained?

The question is only intensified if, as I suggested in the previous chapter, there are no neat "essential natures" that capture the nub of the categories dispositionally delineated by our naive theories. It seems that the most we might extract from the intense examination of a naive theory's consequences—the most that we might gain from armchair reflection—is an organized restatement of the decision procedure for category membership that is already implicit in the theory, an economical summary of our categorical inclinations. After so much elaborate philosophical footwork, what we get is no more than the contents of our own head served back to us on a silver platter.

Similar objections to philosophical analysis can be made without assuming conceptual inductivism or referential reflexivity. We could easily have had other concepts, suggests Machery (2017, chap. 4), developing an argument from Stich (1990). In place of the concept of water, for example, we might have had a related but somewhat different concept that could be called the concept of water*; likewise, in place of the concept of knowledge we might have had the concept of "knowledge*." (Perhaps water* is H_2O and knowledge* is justified true belief.)

Indeed, continues Machery, there is a modest but growing body of evidence pointing to considerable diversity in the human conceptual inventory, both across and within demographic groups. Differing responses to the Kripkean "Gödel" scenario among Westerners and East Asians might be understood as showing that these two groups have different understandings of the workings of reference, one causal-historical and the other more descriptivist. Further, even among Western respondents, there is enough variation in judgments about the case to suggest that thinking about reference might differ from person to person as well as from culture to culture, with a substantial minority of ordinary Americans understanding questions about reference to be inquiring about reference*. (More peculiar things have happened; on November 8, 2016, for example.)

Whether the diversity is psychologically real or merely possible in principle, it raises a question: why should we armchair philosophers invest so great an effort in analyzing knowledge and reference and singular causality when we could be analyzing knowledge* and reference* and causality*? We have no antecedent reason to suppose that our naive, unstarred concepts are better gateways to philosophical knowledge than their starred cousins. Yet we work exclusively on the former. Most ordinary working analysts do so unthinkingly, I presume, but self-conscious analysts owe us a justification for their choice. According to Machery they have only two options: to insist dogmatically that their own concepts are best, or to concede that they have no objective reason to prefer their own concepts while parochially plumping for them all the same.

Either way it seems clueless to be so captivated by our "starter concepts," to be so invested in the consequences—inductive or otherwise—of our naive theories. Why not put that time and effort to work building better theories? At the very least, before commencing analysis we should ask with Sally Haslanger (2000, 33):

> What is the point of having these concepts? What cognitive or practical task do they . . . enable us to accomplish? Are they effective tools to accomplish our (legitimate) purposes; if not, what concepts would serve these purposes better?

Machery poses the same question, as do the ameliorative epistemologists Michael Bishop and J.D. Trout (2005), the conceptual ethicists Alexis Burgess and David Plunkett (2013), and the conceptual engineer Herman Cappelen (2018).

The fullest defense of philosophical analysis as it is practiced today would provide some reassurance that we can reasonably expect analysis of the philosophical categories picked out by our starter concepts to provide us, more or less as effectively as any other mode of investigation, with conclusions that are substantial as well as true. That is the purpose of this and the next two chapters.

12.2 The Stuff of Philosophy

The subject matter of philosophy might be divided into domains: the epistemic, the moral, the mathematical, the material, and so on. (Such a division does not preclude overlap: perhaps the moral or the mathematical facts are a special kind of material fact.) Our everyday philosophical concepts—knowledge, justice, causality—purport to pick out substantial features of these domains; the analysis of the categories ought, in that case, to give us substantive knowledge of the domains. It is on such a posit that the importance of philosophical analysis turns.

I propose to slice the posit in two. On the one side I put the assumption that there are substantive facts to be found in each philosophical domain—that there are epistemic facts worth knowing, moral facts worth knowing, aesthetic facts worth knowing—and that (as this phrasing presupposes) our concepts and words are capable in principle of picking out such facts. On the other side I put the assumption that our philosophical categories correspond to important features of their domains, rather than picking out arbitrary, parochial, trifling, or otherwise uninteresting aspects of the subject matter. Not only are there consequential properties and classes to be found in the epistemic domain, for example, but our human concept of knowledge, in its nature or its membership, corresponds to such a thing—so the assumption goes.

In the following three chapters on the substantiality of philosophical analysis, I commit to the first assumption without any argument. A full and complete vindication of armchair philosophical inquiry would not, of course, take the existence of interesting and important philosophical facts for granted. The significance of each philosophical domain would have to be queried in turn. Is there really a material world? Does it have spatiotemporal structure? Nomological structure? Causal structure? If such structure exists, is it mind-independent? If not mind-independent, is it nevertheless sufficiently consequential—sufficiently objective, or deep, or important—to constitute a worthy endpoint for philosophical inquiry? Then the moral domain: Are there really moral facts? Are they determined by or identical to material facts? If so, are they of philosophical, as opposed to merely anthropological, interest? All such questions must be answered, whether by philosophical analysis, empirical investigation, or meta-philosophical power moves such as an appeal to the reflexivity of reference. But this book will pass right over them; it simply presumes that there are philosophical facts that can and should be known.

That will give me the leverage I need to test the second assumption, that analyses of the concepts with which we start out in life—the pretheoretical concepts of causality and knowledge, the beautiful and the good—are an effective way to learn at least some of the substantive philosophical facts. In so doing, I counter the "negative" experimental philosophers' contention that our pretheoretical concepts are too primitive, too run through with prejudice and falsehood, to get at any truth worthy of our philosophical attention.

To further focus the debate, I will not only assume the existence and importance of the various philosophical domains; I also suppose that the elements of each domain sit in explanatory relations to one another and perhaps to elements of other domains. Within the moral domain, for example, the moral valence of certain acts will be explained by other moral and nonmoral facts. Why is waterboarding criminal suspects wrong? The wrongness is not, presumably, a fundamental moral truth, but rather follows from physical, biological, and psychological aspects of waterboarding along with various ethical precepts. To give an account of the wrongness of waterboarding is to give an explanation—a noncausal explanation—of its wrongness.

Likewise, moral valence may explain some nonmoral facts; the wrongness of waterboarding might, for example, play a part in explaining why a certain government refrains from using it even as a last resort. (Perhaps it might also explain why a different government, rotten to the core, goes out of its way to use it.) Another example, the explanatory structure of the epistemic realm, will be explored at length later in the next chapter.

The assumption of explanatory structure allows the notion of "substantial knowledge" to be given some character: knowledge of a philosophical domain is substantial, I will suppose, to the extent that it elucidates the domain's explanatory structure. Categories are substantial to the extent that they capture explanatorily meaningful distinctions.

The present chapter is almost entirely concerned with the question of substance as it arises in the analysis of basic natural kinds such as water and swan. On the one hand, it might seem obvious that asking "What is water?" is a good way to gain worthwhile chemical knowledge. On the other hand, the negative experimental philosophers' concerns seem to apply even in this case: the category of water is picked out by a concept that is chemically naive; it does not line up in a neat way with chemically sophisticated categories such as H_2O; there are many alternative categories that might do the same job equally well if not better. Surely it is wise to regard our naive chemical concepts only as unavoidable starting points for chemical research, abandoning them as soon as is feasible, rather than making their analysis a supreme philosophical goal?

Adjudicating this argument will provide many valuable insights that I will apply in the next chapter to the question of the value of analyzing the philosophical categories. Not everything that is needed to understand the substantiality of philosophical knowledge will have been provided by these two chapters, however; another important part of the picture is provided in Chapter 14, which concludes the series of three chapters on substance.

12.3   The Homing Instinct of Natural Kind Concepts

A fruit is the seed-bearing structure of a plant—to a botanist. Botanical fruits therefore include acorns, allspice, chili peppers, corn, pumpkins, and of course tomatoes. As the US Supreme Court established in *Nix v. Hedden* (149 U.S. 304 [1893]), however, in the ordinary nontechnical meaning of the term, tomatoes are not fruit but vegetables.[1] Likewise, in the everyday sense acorns are not fruit but nuts—although to a botanist, nuts are strictly a subcategory of fruit (for which reason cashews, pecans, and walnuts are not nuts in the botanical sense). To top off the terminological sundae, bananas are botanically berries but strawberries are not. The ordinary or "culinary" concepts of fruit, nut, and berry certainly have their applications. But their analysis will not tell you much about the biological world.

Some writers have argued that the concept of race, and individual race concepts such as "East Asian" and "Caucasian," have built into them a false essentialist presupposition. Hirschfeld (1996), for example, suggests that concepts of race have the same essentialist structure that the psychological essentialists attribute to the basic natural kind concepts. To understand biologically based human differences, you might usefully conduct DNA analyses or physiognomic studies. But if Hirschfeld is right, then analyzing the con-

---

[1] More exactly, the Supreme Court ruled that the ordinary nontechnical meaning was the relevant one for the interpretation of statutes; both parties to the case accepted that in the everyday sense, tomatoes are not fruit.

cept of race would tell you nothing—apart from turning up a feature that all humans have in common, namely, a deep-rooted false belief.[2]

In the case of fruit, analysis of the naive category is biologically uninformative because the concept is tuned to our culinary interests rather than to the structure of the living world. In the case of race, analysis of the naive category is uninformative because the concept is built around a false belief. On the one hand, the concept is too parochial, on the other too Procrustean, for the corresponding category to serve as an illuminating target for analysis. What reason do we have to think that our philosophical categories are any better?

The question's bite is further sharpened by contemplating the unstinting labor that successful analysis turns out to require, when it is possible at all. Finding a condition that draws the categorical boundary just so between all those tricky edge cases—between avocados and olives, perhaps—might be the stuff of dozens of articles in the *Journal of Kitchen Metaphysics*. That is surely so much wasted effort. Likewise for the time and patience expended on the analysis of knowledge, causality, and agency: has that investment been any wiser than the follies of the culinary ontologists? My aim, in a biologically legitimate nutshell, is to argue that it has.

There is a class of concepts that has an encouraging tendency to hook onto categories that really matter—the natural kind concepts. The ordinary fruit concept may better reflect the needs of the dessert menu than the topology of the tree of life, but folk genus terms like "pineapple" or "apricot" seem for the most part to divide up the world along biologically meaningful boundaries.

That is no coincidence, if causal minimalism or something like it is correct. Basic natural kind concepts are connected to other concepts, according to minimalism, by sets of beliefs that constitute modest biological or chemical theories. As evidence arrives and is applied to these theories through inductive reasoning, their Procrustean assumptions will be refuted and their parochialism diluted and eventually washed away.

---

[2] As you might guess, I am not so sure myself that naive thinking about race is technically essentialist.

The assumption that there exist racial essences, for example, being a belief as ordinary as any other, is subject to empirical refutation. Other presuppositions are equally vulnerable to the dictates of evidence, no matter how central their role in a starter theory. If reasoners seek out evidence and reason correctly, Procrustean conceptual structure need be only a temporary impediment to substantive thought.

The beliefs that constitute a parochial concept might be free of falsehood, but they primarily concern the connections between the corresponding category and properties relevant to certain highly specific human endeavors—as sweetness and taste are prioritized by the need for a satisfying end to the meal. The effect of those priorities will diminish, however, as more information arrives and the beliefs cohere into a theory that paints an ever more complete picture of local explanatory reality. What starts out as highly parochial will be supplemented and deepened over time until something close to a mature scientific theory takes shape.

The learning mechanisms by which this process unfolds have been discussed extensively earlier in the book (Chapters 5, 6, and 9 in particular). Especially important in the present context is the way in which causal minimalism was developed in the treatment of Twin Earth style scenarios (Section 9.6). I proposed there that core causal beliefs are in effect individuated by explanation: each belief connecting a kind to a characteristic property targets a particular (though typically unknown) explanation of that property. As a consequence we will eventually, on receipt of a set of total evidence, classify as belonging under a kind only substances or organisms in which the characteristic properties are each explained in a single way. Because H_2O and X_yZ are transparent for different reasons, for example, they fall under different core causal beliefs of the form *Something about ____ causes transparency* and so get connected ultimately to different basic natural kind terms ("water" and "Twin Earth water" respectively). It is a consequence of this constraint on the core causal beliefs that the theories at the heart of basic natural kind concepts have not only a tendency to truth and impartiality (as opposed to Procrusteanism and parochialism), but on top of that a tendency to line up along explanatory boundaries.

In Chapter 1, I distinguished "closed" concepts, which execute a preordained plan for category delineation, from "open" concepts, which when drawing boundaries between categories take into account the nature of the world into which they find themselves thrown. Inductive concepts that are centered on explanatory beliefs, and causal minimalist basic natural kind concepts in particular, are paradigms of openness. It is through this receptivity that they shed false preconceptions and sideline parochial preoccupations. I conclude the present section with a few qualifications of this idea.

First, I don't claim that every natural kind concept will unerringly anchor itself to a handsome chunk of the explanatory web. Natural kind concepts are not invulnerable to the falsehood of their starter beliefs. Too many in the wrong context, and a concept will fail to pick out any category at all, let alone one that is substantial or objective. But inquiry always carries some risk, and a few failures of reference are tolerable.

Second, I treat the world's explanatory structure as substantial and objective, and so as a worthy aim of inquiry. But what if the enterprise of explanation is itself colored by local interests? Then the schema inscribed on the world by our natural kind concepts, though it may align with explanatory boundaries, would be no more than an expression of a higher parochialism.

While I am keenly alive to this possibility, I do not think it is necessary to wrestle with it here. Scientific knowledge is structured by explanatory concerns. Think of this and the next two chapters as attempting to elevate armchair philosophical inquiry to the same level, with respect to substantiality and objectivity, as scientific inquiry. That is good enough for now.

Third, there appear to be counterexamples to the thesis that causal minimalist concepts converge inexorably on objective explanatory structures. I gave one above: the everyday or "culinary" concept of fruit, which has clearly not escaped the orbit of the dinner table; likewise, the everyday concepts of nut and berry. Because they taxonomize above the genus level, they are not technically basic natural kind concepts, but an inductivist will want to understand them as built around theories of some sort. Why do these theories stay parochial?

One possible response is that the fruit concept has something other than a minimalist structure—that it was not built from core causal beliefs, but rather from (say) a functional definition. I doubt this is correct; certainly it is uncomfortably ad hoc. Let me give a different answer. The starter fruit concept, I allow, was most likely built from core causal beliefs arranged in the characteristic starburst, and was therefore subject to the same explanatory pressures as the folk genus concepts. But various external factors prevented these pressures from having their full, customary effect. The fruit concept, like the nut and berry concepts (and the bug concept discussed in Section 10.4) has bifurcated. Along one branch it has acted inductively, converging on a biologically significant category. This has given us the botanical concepts of fruit, berry, nut, and bug.[3] In doing so, it has created space for another concept to reject the overtures of science and to occupy what is in effect a social role, in the following way.

After bifurcation, the ordinary fruit concept, I suggest, continues to be a causal minimalist concept with substantial explanatory content, such as a core causal belief to the effect that something about fruit makes it sweet. It is because we remain committed to this and similar causal connections that we are unwilling to classify cucumbers and peanuts as fruit. These causal commitments are unstable because fruit in the ordinary sense is not an explanatorily homogeneous category. But the concept is exceptionally useful just the way it is, and scientific pressure to reform can be shrugged off, thanks to the botanists' efforts, with the remark: I don't mean fruit in *that* sense. So the concept persists in its current form, and thanks to that persistence, refers to just what we take it to refer to (except perhaps in a few uncertain cases such as the perennially problematic rhubarb).

To summarize the argument so far: the theories that make up basic natural kind concepts have a tendency to develop, under the dual stewardship of empirical evidence and reflection, into rich, cohesive, accurate, explanatory theories—chemical theories for substance concepts, biological theories for folk genus concepts. Kind concepts will tend to refer from the start, then, to categories that play an important role in the world's explanatory economy, rather than to features of the explanatory terrain that are only of parochial interest or to nothing at all.

Can we expect the same convergence in the starter theories that make up philosophical concepts? That, it turns out, will depend. For the purposes of vindicating philosophical analysis, I will divide the philosophical categories, oversimplifying considerably, into two classes. The starter concepts in both classes take the form of explanatory theories. For different reasons,

---

[3] I do not see these concepts as having been created, then, by a conscious act of definition. Few scientific concepts are; even novel theoretical terms are typically introduced nondefinitionally (Strevens 2012c).

these theories typically provide a good basis for inquiry into something of substance, something worth caring about.

In the case of one class—the class that is more like the natural kinds, and which includes categories such as space or Darwinian fitness whose investigation is more sensitive to empirical matters—the reason has to do with convergence shaped by new information obtained through the senses, through reflection, and perhaps in other ways. As with the basic natural kinds, I will suggest in Chapter 13, the convergence washes away the parochial and repairs the Procrustean elements of starter theories.

In the case of the other class—which I surmise includes many of the philosophical categories that we associate most strongly with armchair analysis, such as knowledge—the reason is different; it is laid out at length in Chapter 14.

By one or the other of these two routes, then, it can be seen that most philosophical concepts consist of theories suitable for substantial inquiry. But to vindicate analysis, further work is necessary. Analysis does something rather specific with a theory: it uses it to answer a question about the essential nature of the corresponding category. Even if you accept the substantiality of a theory, you might doubt that this is a promising way to put it to work.

I have in mind two worries in particular, each of which calls into question the value of analyzing basic natural kinds as much as it does the value of analyzing philosophical categories. The first worry picks up on my suggestion that both basic natural kinds and philosophical categories are likely, in many instances, to lack essential natures. The second worry asks whether the effort analysts put into making judgments about peculiar or recondite edge cases of category membership is really worth their while. These issues will be tackled in the next (and final) two sections of this chapter, Sections 12.4 and 12.5.

A third worry concerns the value of analyzing concepts that turn out to be empty—concepts that fail to refer to a category (or perhaps, that refer to an empty category). I take up this issue in the next chapter, in Section 13.3.

12.4 Analyzing Natureless Categories

Ideally, analysis of a basic natural kind will reveal an essential nature that plays a significant role in the world's causal-explanatory structure. If my conclusions in Chapter 11 are correct, however, then many basic natural kinds have no essential nature. Analysis of such a kind, you might think, will surely lead to disappointment rather than explanatory enlightenment. In what follows, I develop this worry into an argument against the value of analysis—an argument that I will then attempt to undercut.

Let me begin with the case of water. In a naive theory of water, waterhood plays a central inferential and explanatory role (Figure 6.1). Various things about water—the naive reasoner knows not what—are represented as causing each of that substance's putative characteristic properties: transparency, potability, conductivity, and so on; or, to say the same thing more formally, waterhood is represented as entangled with the causes of its putative characteristic properties. By classifying or seeking to classify a specimen as water, then, the reasoner is able to make a number of useful, causally sophisticated predictions. The notion of water is at this stage cognitively indispensable.

As the evidence comes in, the reasoner begins to learn what is actually causing the characteristic properties. If the world were, in a strong sense, an essentialist place, the cause might be in each case the property of waterhood itself. A sophisticated scientific theory of water would then accord waterhood the same central role as it has in the naive theory, with the arrows that earlier meant "is entangled with a cause of" upgraded to the simpler and stronger "is a cause of." In a world like ours, however, the cause of each characteristic property is something other than waterhood. The naive theory of water is shown to be more or less true, but at the same time it is rendered redundant by more advanced theories about the direct causes of the properties in question: chemical properties of H_2O, of the solutes typically found in H_2O, and so on. The category of waterhood remains a useful cognitive organizer, but the explanatory theories underlying the behavior of water seem not to require the notion of waterhood at all. The analysis of water, then, may turn out to be a meager explanatory achievement, and therefore a commensurately meager contribution to scientific knowledge.

I would wager that most basic natural kinds will, like water and swan, turn out to have no essential natures, and that their explanatory significance in the physical and biological domains will turn out to go by way of relations of entanglement rather than causation. As we learn more about the world, then, the basic natural kinds will tend to be pushed into the causal-explanatory background by new categories that are directly explanatory—namely, the true causes of the kinds' characteristic properties, revealed gradually by scientific inquiry. The natural kinds do a useful job, embracing clusters of properties that do the causing, but they do not cause anything themselves. At the end of inquiry, when our causal knowledge is complete, the explanatory enterprise will have no further use for the natural kinds, much as Quine (1969) suggested. Our explanatory discourse will replace all talk about folk substances and folk genera with talk about molecular structure, developmental genetics, metabolic pathways. And so, though the basic natural kinds have an important organizational role to play, their analysis will turn out—often enough—to be of little scientific importance.

Is that good enough reason to give up on the enterprise of analyzing the basic natural kinds? I say no, because even when a kind has no essential nature to discover, its analysis will tend to lead, in a relatively efficient way, to copious and significant knowledge of the underlying causal-explanatory structures. The analysis of basic natural kinds can be justified, then, even when that analysis is doomed to fail, by its collateral benefits.

It may seem peculiar, if not positively perverse, to encourage investigators to pursue these benefits in so indirect a way. If information about the causes is what we want, why not investigate causal structure directly, rather than analyzing basic natural kinds whose connection to the causal network is only indirect?

My response is that we do not know in advance whether our pretheoretical categories have explanatorily important essential natures or not. The first chemists rightly sought to analyze water—they posed the question "What kind of thing is water?"—because waterhood might well have played a causally central role in the chemical domain. If that had turned out to be the case, they would have gotten exactly what they wanted. Of course it did not: water has a theoretically uninteresting nature or it has no nature at all, and either way it participates in the explanatory economy only via its entanglement with the true causes of chemical appearances and behaviors. Yet the pursuit of the question was, for a time, immensely fruitful. That chemists once sought the nature of water is no cause for regret whatsoever.

At this point in time, the value of the water project is more or less exhausted. However much philosophical souls might like to play around with the analysis of waterhood, there is no serious prospect of learning more chemistry by doing so, principally because we know so much already. That is less obviously true of the analysis of the folk genera, however, and surely false of the case that ultimately concerns me, the philosophical categories. The biological taxonomists, the ethicists, and the epistemologists are, then, much like the early chemists, and they should pursue the analysis of their categories, the attempt to answer "What is . . . ?" questions, in the same spirit. If there are essential natures to be discovered, superb. If not, the investigation stands to reveal a great deal about the world's explanatory structure all the same.

12.5 Thinking on the Edge

Not so fast, says the skeptic. The analysis of natureless categories might indeed lead to explanatory knowledge, but it is an absurdly inefficient way to get there, at least past a certain point that we long ago left behind. As armchair investigators know from long experience, the advanced pursuit of analysis directs great quantities of intellectual energy, talent, and time toward the consideration of peculiar edge cases: coffee-flavored washing-up water, factory-assembled swan egg duplicates, real barns in fake-barn country, preempted double preventers, and so on. Such efforts make sense in the pursuit of knowledge of essential natures, but they seem almost if not entirely worthless if the true goal of the investigation is knowledge of underlying causal-explanatory structure—knowledge of the causes of water's transparency, swans' red beaks, and the rest. A would-be analyst of a natureless category might well stumble upon a few well-lit explanatory clearings in the course of their endeavors, but they will spend the majority of their days pushing through the thick, dark brush of tangled counterexamples, ad hoc revisions, and obscure definitions. It looks to be a gratuitously painful path to discovery, if what stands to be discovered are not essential natures but causal structures.

This gloomy line of thought, I hope to establish, is far too pessimistic. The meticulous intellectual dissection of bizarre scenarios and remote possibilities is not merely an effortful distraction: it leads, as I will now show, directly to explanatory knowledge.

To understand how the analysis of edge cases aids the acquisition of knowledge about causal-explanatory structure, we must explore a new subject matter: the cognitive function of explanatory thinking, or in other words, the question of what ordinary thinkers stand to gain from dwelling on questions about what explains what, with special emphasis on the role of the entanglement relation. That rather long but I think intrinsically interesting journey begins here. Neither edge cases nor analysis will appear right away, but they will be along presently.

Why We Are So Clever

Swans have two legs. Like almost every generalization we encounter in everyday life, this one has exceptions: hopping down the side of the pond, sooner or later, comes a one-legged swan.

If you parse the generalization as an irreducibly statistical claim—*Most swans have two legs*—you will not get much help in anticipating the exceptions. You have to accept them, when they come along, like a gambler swallowing a streak of bad luck or a peasant enduring inexplicable visitations of wrath from an angry god.

It is possible, however, to be far savvier about swans. If you understand the generalization as making a claim about the consequences of a certain causal mechanism, and you have some independent knowledge of the mechanism—even just a little—you can leverage your knowledge of the world to make intelligent surmises about the circumstances under which exceptions will arise. Knowing that swans naturally grow two legs, for example, you can infer that one-leggedness in swans is likely to be the result of an accident. Older swans are more likely to have suffered such an accident, so older swans are more likely to be one-legged, and vice versa.

Thinking not only about effects, then, but about how those effects are to be explained, better enables you to anticipate exceptions or to make inferences from known exceptions to further conclusions. To improve our thinking in this way is one great function of causal cognition, and more generally, I would hazard, of explanatory cognition. Thinking explanatorily makes you smarter.

My first business in this section is to show that what I have said above applies, in the case of basic natural kinds such as water and swan, to entanglement: the function of representing a relation between kind and characteristic property as one of entanglement rather than, say, statistical correlation is to sophisticate the thinker's treatment of exceptions.

The second stage then unfolds as follows: Doing analysis involves much thinking about category membership. For categories, such as basic natural kinds, whose cognitive significance inheres in various beliefs about their role in explanatory structure, membership judgments frequently involve inference to the best explanation and related moves. In particular, deciding where to draw the line between exceptions and near-exceptions—the analyst's foremost task—requires the application of fine-grained explanatory knowledge. Even if the relevant explanatory relations are entanglements, such knowledge will concern a great deal of underlying explanatory structure. Case judgments about basic natural kinds, then—even kinds with no essential natures—will enrich the judge's knowledge of causal structure, systematizing what they already have and motivating them to seek out what they do not.

The Nature of Entanglement

As you have seen, a core causal belief such as *There is something about swans that causes whiteness* spans two relations: of some typically unknown mechanism P, it says that swanhood is entangled with P and that P causes whiteness. I need to tell you more about the entanglement side of such beliefs.

If swanhood is entangled with P, I said in Section 6.1, then there is a counterfactually robust tendency for swans to have P. This tendency manifests itself in three ways. First, most or all swans have P. Second, in a range of counterfactual circumstances—in slightly changed environments, with slightly changed diets, or slightly different upbringings—those swans would for the most part also have had P. Third, in a range of counterfactual circumstances that would have brought about the creation of swans that do not actually exist, such as counterfactual matings between actual swans, those nonactual swans would also, for the most part, have had P.

Entanglement is, because robustness is, a matter of degree. The wider the range of counterfactual circumstances in which P comes along with swanhood, the stronger the entanglement. I will assume that core causal beliefs represent the entanglement as "strong enough"; with that supposition, I put the graded nature of entanglement aside.

Entanglement is asymmetric. That P usually comes along with swanhood does not mean that swanhood usually comes along with P—that most organisms with P are swans. Perhaps the causes of whiteness in swans are the same as in many other waterfowl.

There are many different reasons that a property F might become entangled with another property P. There might be a mechanism by which F-ness causes P-hood. Or the usual cause of F-ness might also cause P-hood. Or P-hood might be metaphysically constitutive of F-ness, or for that matter the two properties might be identical. Or P's accompanying F might be mandated by a metaphysical law, or a noncausal law of nature, or a moral law, or anything else sufficient for a robust connection.

Most important of all, perhaps, are cases where for some quite contingent (though not necessarily chance) reason, the Fs come to have P at some stage, and where some mechanism then passes P reliably and robustly to later instantiations of F. A founder population of ravens, for example, may have for whatever reason tended to blackness; from that point on, blackness might have been reliably passed on to later generations due to some mix of selection for species recognition, sexual selection, and mild genetic or phylogenetic "inertia." Or consider the earth's tectonic structure: whatever its causes, it persists due to a kind of long-term "geological inertia," perpetuating an entanglement between "Earth-hood" (a property with a single planetary instance though many temporal stages) and various structural properties of the crust.[4]

Robust association is only one half of entanglement; the other half comes in several pieces that work together to make the reasoner who thinks in terms of entanglement cleverer than the purely statistical reasoner. First, as the "something about" formulation suggests, if F is entangled with P then P must be an intrinsic property of Fs. Swans' entangled whiteness-causers must, for example, be intrinsic properties of swans. (That is not the same thing, note, as the whiteness-causers being an intrinsic part of swanhood.)

---

[4] These ideas are further explored, though without explicit reference to entanglement, in Strevens (2008b).

Second, in order for F to be entangled with P, there must be no property F^* more general than F—a property F^* such that a thing's having F necessitates its having F^*—that is also entangled with P. The entanglement of F and P requires, then, that F is the most general property of the Fs for which the tendency to P-hood exists. If all the *Anatidae* share a mechanism for growing webbing between the toes, then it is not swanhood but membership of the anatid family as a whole (or perhaps even some larger taxon) that is entangled with that mechanism.

A third condition on entanglement limits this drive to generality: F is entangled with P only if in almost all cases where P comes along with F, it does so for the same reason. If the entanglement exists because of some causal mechanism, it must be the same mechanism in each case. If it exists because of a moral law, it must be the same moral law in each case. There must be, then, a certain homogeneity to the robust connection.

Much more would have to be said about the individuation of reasons for entanglement in order to give this condition substance. For my purposes here, it is perhaps sufficient to stipulate that reasons for entanglement are to be individuated explanatorily: in order for P to come along with F for the same reason in almost all cases, the explanation of P's coming along with F must be the same in almost all cases.

It is this connection between entanglement and explanation that gives entanglement a special role in the cognitive economy, as I will now show.

The Uses of Entanglement

A typical swan contains a cluster of overlapping but distinct causal mechanisms responsible for its characteristic appearances and behaviors. In addition to representing, or at least representing the existence of, those mechanisms, as swan-savvy reasoners we should represent the fact that they come in clusters. It is this additional information that allows us, when conditions are right, to infer from a bird's white feathers and red beak that it is likely an accomplished trumpeteer.

There are various ways this clustering might in principle be represented and attached to the mental term "swan."[5] An essentialist theory would posit

---

[5] A big question: why do we need the "swan" representation in the first place? Why not simply represent a cluster of mechanisms, with no middle term "swan" sitting at the center of things,

12.5. THINKING ON THE EDGE

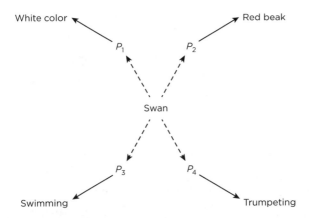

FIGURE 12.1. A causal minimalist theory of swanhood representing the two-step nature of the core causal beliefs. Dashed arrows represent entanglement; solid arrows represent causation.

a single property—the essence of swanhood—that is causally responsible for the appearance of each of the clustered mechanisms in swans. A statistical theory—in effect, an adaptation of the prototype theory of concepts—would represent the clustering as a statistical fact without attempting to explain it, and would define swanhood in terms of the clusters: "x is a swan just in case it has such and such a cluster of mechanisms" (give or take, perhaps, a few). Finally, a minimalist theory represents the clustering by showing swanhood as entangled with the mechanisms.

To make the minimalist strategy explicit, Figure 12.1 expands the causal minimalist theory of swanhood shown in Figure 6.1 so as to show the entanglements represented within the core causal beliefs. The solid arrows represent causal relations; the dashed arrows represent relations of entanglement. The P_is represent the typically unknown causes of the various putative characteristic properties. The representation allows but does not demand that these causes are identical. (Biological sophisticates will know that they are not identical, though they overlap to a considerable degree.)

---

where it must be visited on every inferential journey around the theory? The short answer: to give entanglement relations something to hang on to. The long answer, while rewarding, must wait until another time.

Why should the human mind represent the clustering of the mechanisms in this rather elaborate fashion rather than plumping for the simpler essentialist or statistical representations of clustering? If you were building the human mind, why might you choose to do things the minimalist way rather than in one of the other two ways?

The essentialist theory, as I have already argued, for all its virtues makes an unnecessary and quite probably false assumption: that there is a single property at the center of the swanhood nexus serving as the ultimate cause of everything that is characteristic about swans. Though it is of course quite possible that our naive theories make unnecessary false assumptions in aid of some greater inferential good (Section 8.4), I see no case for engineering an essentialist rather than a minimalist conceptual structure; let me therefore put the essentialist hypothesis aside. What difference, then, does it make to go with entanglement rather than the statistical option?

Suppose you entertain a minimalist theory of swans that attributes to the swans among other things a mechanism for producing white feathers—P_1 in Figure 12.1. You have, in front of you, a bird that you have reason to believe possesses this swan whiteness mechanism. Normally, this would boost your confidence that the bird in front of you was a swan, and thus it would boost your confidence that the bird had the swimming mechanism P_3 and so that it was a swimmer. But suppose also that you know that the reason for the whiteness mechanism's presence in this particular specimen is not the reason that it is typically present in swans. Perhaps, for example, you know that this bird's coloration mechanism has been degraded by disease or malnutrition or a genetic anomaly, resulting by chance in something functionally identical to the swan mechanism (in which whiteness is in fact a consequence of a lack of pigmentation, exposing the natural color of keratin, the material from which feathers are made). That information will, other things being equal, undermine your inference to swanhood, and so your inference to the presence of the swimming mechanism. You will not (again ceteris paribus) think the bird any more likely than before to be a swimmer.

Or to give another example, you might encounter a powder that tastes of almonds due to the presence of the same chemical that causes almond flour to taste of almonds. If you know that this flavoring agent is there for a nonbiological reason, then though the taste is produced in the same way as in almonds, you should not regard it as evidence that the substance is almond flour.

The form of the above story should be familiar. It is the same pattern of inference that I introduced, on behalf of the essentialist and the minimalist, to explain why Keil's subjects refuse to infer, from a cross-dressing raccoon's appearance and behavior, that it is a skunk. To be able to refuse such inferences has the same advantages in both cases: if a cluster of properties go together for some reason—if there is an explanation for their clustering—then you should nevertheless hesitate to infer the presence of the one from the presence of the others in those cases where you know that the relevant parts of the clustering explanation do not apply. Follow this principle, and you become in your inferences a sophisticated applier of hedging conditions, rather than a mere statistician: you not only expect the occasional exception to a correlation; you can in certain cases—those where you have some knowledge of the workings of the relevant explanatory relations—see the exceptions coming and so avoid falling into error.

One further situation in which this sort of reasoning makes itself manifest is the tendency of many biological taxonomists to regard a swan hatched from an accidentally created molecular duplicate of a swan egg as something other than the real thing—to refuse to categorize it as a swan (Section 10.3). The "swamp swan" has the same causal mechanisms explaining its characteristic properties as a real swan, but it has them for the wrong reasons: they are due to replication, not reproduction. The inference from possession of the mechanism cluster to swanhood is therefore blocked. (Divided opinions concerning this case were explained in Section 10.7.)

On such a picture of its cognitive role, entanglement functions as a stand-in for a complicated and typically unknown causal-explanatory network, covering the various evolutionary and developmental pathways that together comprise "the reasons that swans have the characteristic property–causing mechanisms that they do." That seems like a sensible piece of cognitive engineering: it enables us to think in sophisticated ways about the ramifications of the causal-explanatory network swirling around swanhood without requiring a vast and complex set of causal placeholders that most reasoners—whose knowledge of the network is typically piecemeal, shallow, and largely negative—will never be able to fill in.[6]

---

[6] Strevens (2007) argues that representations of dyadic causal relations—"arrows" connecting properties as cause and effect—function in a similar way, standing in for complex causal

The Fruits of Analysis

When using core causal beliefs to judge that a specimen belongs to a basic natural kind, we are in effect judging that the specimen's characteristic properties are explained "in the right way." To be an expert categorizer, then, you must have considerable knowledge of the difference between "the right way" and "the wrong way" when it comes to explanation—knowledge, that is, not only of a kind's characteristic properties, but of the way in which those properties are, in a kind member, characteristically explained, of how the explainers themselves (e.g., proximal mechanisms for plumage coloration) are characteristically explained, and so on all the way down. (Entanglement makes space for indefinitely many links in the explanatory chain.)

An endeavor that exercises your categorical skills, then, also stands to enhance your explanatory skills—in particular, your knowledge of and ability to reason about causal structure. Philosophical analysis, with its focus on the classification of especially complex or questionable specimens, is one such endeavor. Thus, philosophical analysis even of a natureless category is apt to bring explanatory expertise.

This is much the lesser part, however, of the reason that philosophical analysts become experts in explanatory structure. The rest stems from what they do with their case judgments once they have made them: they formulate and test hypotheses about essential natures.

Specimens are counted as category members or not, as I have said, in virtue of whether their properties are explained in the right way or not. A certain transparent, potable substance will be classified as nonwater because it contains no H_2O, where H_2O is deemed essential to the explanation of transparency and so on in water. A duck genetically engineered to resemble a swan will be classified as a nonswan because its cygnid morphology is due to mechanisms that are present for explanatory reasons—having to do, ultimately, with artificial genetic manipulation—that differ from the reasons that the corresponding mechanisms are present in swans.

---

mechanisms in such a way that thinkers can apply their limited knowledge of a mechanism in sophisticated ways without having to represent explicitly a specification of the mechanism.

These rationales, I proposed in Section 4.3, are at least partly accessible to introspection. So we will tend to draw upon them in constructing our hypotheses about the apodictic criteria for category membership—about essential natures. We consequently find it natural to consider very seriously a "microstructural" approach to the metaphysics of basic natural kinds: what matters to being water or to being a swan is the low-level cause of the characteristic properties, that is, the molecular configuration—of hydrogen and oxygen or of DNA—that constitutes or creates the mechanisms for the manifestation of those properties. In the biological case, we also take very seriously a metaphysical approach that focuses on the historical facts that explain microstructure, that is, facts having to do with reproduction, lineage, and evolution.

To investigate the essential nature of a category, then, is among other things to attempt to make explicit those elements of the world's explanatory structure that are relevant to accounting for the characteristic properties of the category's members. The analyst typically does not contemplate explanatory structure explicitly, and they may of course devote a proportion of their attention to hypotheses about natures that have only a passing connection to explanatory matters, but knowingly or not, they will also spend a great deal of time thinking explanatorily.

They may eventually abandon hypotheses about a kind's essential nature built around the "right" explanations for its characteristic properties, concluding that the kind lacks an essential nature. At that point, they should give up on the analysis and the complex edge cases along with it, as chemists have largely moved on from worrying about the proper definition of "water." But they will not regard their efforts as wasted. In the course of their failed analysis they will have become experts in something very much worth caring about, the causal-explanatory structure in the kind's vicinity. That in itself will have made the analytic inquiry worthwhile.

CHAPTER 13

Substance: Philosophical Categories

13.1 Beyond Basic Natural Kinds

My story about substance has so far focused on the analysis of basic natural kinds, turning on their corresponding concepts' characteristic causal starburst structure. In order to vindicate the analysis of philosophical categories, I will now take a broader perspective, open to concepts with other explanatory structures and relations.

Every domain susceptible to philosophical investigation—I am tempted to say, every domain of human knowledge—has an explanatory structure, I assumed in Section 12.2. The explanatory relations that make up this structure differ from domain to domain. In the material domain, causation is the principal such relation.[1] Outside the physical realm, I suggest (following a long philosophical tradition; e.g., Kim (1994)), there is explanatory relevance wherever there is the right kind of dependence and difference-making. I call these difference-making dependences the *primary* explanatory relations. Mathematical difference-makers are primary explainers of the mathematical facts to which they make a difference. Metaphysical difference-makers are primary explainers of the metaphysical facts

---

[1] Many philosophers would point to metaphysical dependence as another important explanatory relation, even within the confines of scientific thought (Strevens 2008a, chap 7).

to which they make a difference. Moral difference-makers are primary explainers of the moral facts to which they make a difference.

What kind of explanatory relation, then, is not primary? In Strevens (2008a) I argue that entanglement is an explanatory relation in a derivative sense: a property that is entangled with a cause is thereby explanatorily relevant to the cause's effects (though not to the cause itself). Consequently, a normal swan's being a swan is relevant to explaining its whiteness, a normal raven's being a raven is relevant to explaining its blackness, a glass of water's waterhood is relevant to explaining its transparency, and so on. (Prasada and Dillingham [2006] provide experimental evidence that ordinary reasoners go along with these judgments of relevance.) Indeed, the antecedents of most high-level laws and other explanatory regularities in the special sciences are explanatorily relevant to their consequents, I argue in Strevens (2014), in virtue of their being entanglers—in virtue of their being entangled with the causes of the consequents—rather than their being causes themselves.

Entanglement plays the same role, I propose, in every explanatory domain: a property entangled with a primary explainer is relevant to whatever that primary explainer explains, whether the explanation is causal, mathematical, or moral. In virtue of this fact, I call entanglement a secondary explanatory relation.

For the purposes of this book, you need not go along with my view that an entangler is a secondary explainer. What is important is that entanglement functions inferentially in much the same way as primary explanatory relations such as causation. In the case of a causal link, you can infer from effect to cause provided you have no reason to think that the effect is the result of some mechanism other than the mechanism represented by the link. In the case of an entangling link, you can infer from entanglee to entangler provided you have no reason to think that the entanglee is present for some reason other than the reason underlying the link.

A secondary explainer that is entangled with more than one primary explainer can, then, organize those explainers for explanatory and inferential purposes in the same way that the basic natural kind categories organize the causes of the kinds' observable properties. It may be in many cases useful, then, to build a theory with the same starburst explanatory topology as the basic natural kind concepts—that is, the topology shown in Figure 13.1.

13.1. BEYOND BASIC NATURAL KINDS

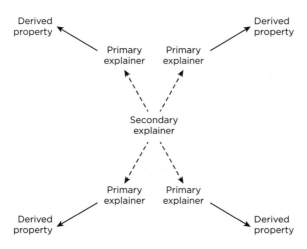

FIGURE 13.1. The starburst topology of basic natural kind concepts generalized. Dashed arrows represent entanglement; solid arrows represent some primary explanatory relation of difference-making dependence.

I call the properties at the periphery of the starburst—for the swan, whiteness, trumpeting, and so on—the theory's *derived properties*. The term "derived" is not an articulation of a property's place in the order of being but merely of its place in the fragment of the explanatory web putatively represented by the theory. In the case of the basic natural kinds, the derived properties are the most palpable properties: a specimen's observable appearances and behaviors. They are epistemically prior (in at least some senses) and explanatorily posterior. That may or may not be true generally. But in any case, you can see that a theory with the starburst topology imposes a certain structure on inference from one derived property to another, traveling up the primary and then the secondary explanatory links to the central, organizing concept, then down other links of the same sort. The purpose of this long and indirect inferential journey is to put certain sophisticated constraints on the reasoning in question: traverse a link only if you have some reason to think that the explainer in question—a mechanism or other difference-maker, a reason for entanglement—is or was at work in the case in question, or at least no reason to think otherwise.[2]

---

[2] A visit to the organizing concept is not always necessary: direct causal links may be represented between various of an organism's physical features, for example.

Imposing such constraints by using the explanatory topology shown in Figure 13.1 has a side effect: you must represent a central secondary explainer. In order to secure a certain kind of inferential connectivity, then, you add a new concept to your mental inventory. So concepts of basic natural kinds and of many other categories, even philosophical kinds, find their way into thought.

If a category is a primary explainer—if, when all the evidence comes in, it is revealed to be a cause or to be an active part of some other primary explanatory network, such as the network of nomological dependence, or mathematical dependence, or moral dependence—then I call it, naturally, a *primary category*. If a category is not itself primary, but it is entangled with primary categories, then it is a *secondary category*.

A few remarks. First, some categories are neither primary nor secondary. They may exist outside of the explanatory order, or they may be related to it without themselves being explainers.

Second, the definition of a secondary category does not merely require that all a category's *known* explanatory relations are entanglements. It requires that all explanatory relations, known or not, go by way of entanglement. Equivalently, then, it requires that on receipt of a total set of evidence, competently handled, the corresponding concept would represent the category participating only in entanglements and incidental nonexplanatory relations.[3]

Third, a category's status as primary or secondary is relative to an explanatory domain. Perhaps every property is a primary explainer in the domain of deductive logic, because a thing's being F, no matter what F may be, explains why it is either F or G.

There are explanatory topologies other than the starburst. In these alternative topologies, primary and secondary explanatory relations will play the same ultimate role as in the starburst, constraining inference between correlated properties according to rules that are far more sophisticated than those warranted by a merely statistical connection. Many philosophical concepts hypothesize such structures for their corresponding categories, I surmise; the philosophical concept of knowledge, to be investigated in Section 13.4, will furnish an example.

---

[3] The equivalence obtains, of course, only if all explanatory relations are knowable, a supposition that I will not question here.

13.2 The Substantiality of Philosophical Analysis

Concepts of philosophical categories consist of ordinary beliefs about those categories' place in the apposite explanatory structure, either as primary explainers, secondary explainers, or in some other role. Making only this assumption, I hope to show that the analysis of philosophical categories tends to produce substantial knowledge.

When vindicating analysis of the basic natural kinds in the previous chapter, I considered three reasons to doubt the value of analyzing some apparently significant category:

1. Our starter concept for the category is likely run through with parochialism, triviality, and falsehood. It constitutes a faulty theory, an unsuitable foundation for substantive investigation.
2. The category may well lack an essential nature—yet determining essential natures is the foremost aim of analysis.
3. Analysis pours tremendous resources into the consideration of bizarre edge cases whose only relevance is to determining essential natures—which may not in fact exist.

My responses to these skeptical concerns were laid out, with one significant omission, in the treatment of basic natural kinds. In dismissing them as challenges to the interest of philosophical analysis, I can therefore be rather brief.

The first skeptical argument casts doubt on starter theories. Referential reflexivity assures us that our initial set of beliefs concerning a philosophical category is a good starting point for inquiry concerning that category. But is the category itself an interesting feature of the underlying explanatory domain? Equivalently, does the starter theory point toward interesting features of the domain?

There are two ways to give a positive answer to such a question. You might argue that the starter theory already articulates some substantial truth about the domain. Or you might argue that the starter theory will with time come to do so—that even if it is itself flawed, it will upon further observation and reflection converge on a theory that articulates substantial truths.

I took the latter path when vindicating the analysis of the basic natural kinds. Empirical investigation, I said, would eliminate falsehoods from our starter theories of the kinds while adding a sufficiently diverse array of truths about primary explanatory structure to alleviate any initial parochialism. The likely end product is a rich, deep, accurate theory of a certain chemical or biological subject matter.

Our starter theories of at least some philosophical categories stand to be improved in the same way. Scientific discoveries about the nature of space, for example, will set our metaphysics of space on the right track. Discoveries in formal logic—think of Gödelian incompleteness in particular—will guide our theorizing about the foundations of mathematics. Moral reflection might bring our ethical precepts closer to the truth, perhaps by "expanding the circle" (Singer 1981).

In general, because our philosophical concepts are composed of ordinary beliefs, they gracefully accept any amount of correction—by contrast with definitions or other stipulations, which stand fast no matter what. Provided we have some source of relevant information, then, it seems that we can escape the traps of prejudice and error.

Yet some of the great analytic projects in philosophy fit this optimistic story rather awkwardly. Our starter theories of knowledge or of singular causation might be revisable in principle, but they change only slowly and perhaps only around the margins. Might it not be that the basic natural kind concepts' supreme openness to the world is essential to their ability to home in on objectively significant elements of the explanatory web? And so that the relatively empirically imperturbable philosophical concepts are not nearly so nimble, not nearly so fast to cast off their parochial and Procrustean ways?

The question is especially acute for those philosophical categories for which the action, when determining essential natures, unfolds almost wholly in the armchair. Heavy reliance on pure thought suggests that the corresponding concepts are largely closed: they are dies stamping their preconceptions onto the world, rather than learning from it what really matters.

There is indeed a major difference between the natural kind concepts and the concepts of many of the best-loved, most analyzed philosophical categories. Like the natural kind concepts, the philosophical concepts

13.2. THE SUBSTANTIALITY OF PHILOSOPHICAL ANALYSIS 257

are inductive and take the form, throughout their existence, of theories of primary explanatory structure—but explanatory theories of a rather special sort. It is this specialness that explains their susceptibility to armchair analysis, and indeed, that explains the "case certainty" that at least some analyzers experience when making judgments about certain scenarios.

The specialness also calls for a somewhat different story about substance, that is, about the reasons we have for expecting our inquiry into the philosophical categories to be an investigation of something truly worth the effort expended. As an alternative to convergence, I suggested above, the would-be vindicator of analysis might argue that the starter theories for some categories spell out, from the very beginning, important facts about the primary explanatory structure of the underlying domain. That is close to the move I intend to make: I will say that the starter theories for philosophical categories such as knowledge make claims that would be interesting and substantial if true, but whose falsehood would also be interesting and substantial. Whichever way the starter theory's posits come out, then, we will have learned something important about the world.

In principle, the development of that idea belongs right here. But it is long and complex, so I have put it in a chapter of its own, Chapter 14.

In addressing objection (1) above, then, my investigation splits into two parts, one going by way of convergence and the other by the way I have just described. In confronting objections (2) and (3), the paths rejoin: I say the same thing about the analysis of space and the analysis of knowledge.

On to the latter two objections. Suppose we undertake to analyze our pretheoretical categories in some domain—chemistry, biology, epistemology, metaphysics. It is quite possible and perhaps even rather likely, my Chapter 11 skepticism about essential natures implies, that these categories' boundaries will fail to coincide exactly with primary explanatory structure, which is to say that their membership will not be determined by possession or otherwise of a primary explanatory property. The analytical project makes sense, then, only if there is some intellectual benefit to be obtained even if the targets of analysis are not primary explainers. Although it is unreasonable to suppose that the majority of our pretheoretical categories are primary explainers, it is reasonable for a conceptual inductivist to think that they are at least, for the most part, secondary explainers—secondary categories. To vindicate the analysis of everyday categories, then, whether in science or philosophy, it suffices to show that there is substantive knowledge to be gained from the analysis of secondary categories.

Entanglement, or secondary explanation, has an important cognitive role to play in thinking about biology, chemistry, and just about any other explanatory domain, including the philosophical domains. That role, however, appears to be wholly utilitarian and ephemeral: it is to represent the co-occurrence of, and to stand in temporarily for, unknown primary explanatory structures such as causal mechanisms or relations of moral dependence. Secondary categories are cognitive tools; they are not intrinsically interesting features of the world, but devices for organizing our knowledge of such features—for organizing our knowledge, that is, of the world's primary explainers. It follows that the analysis of secondary categories produces substantive knowledge only if it produces significant knowledge of primary structure.

In the vicinity of a secondary category, entanglement has two cognitive functions. On the one hand, it is as I have said an explanatory stand-in, a placeholder for some primary explanatory relation, or for a chain of such relations. On the other hand, nestled closer to the categorical property—waterhood, swanhood, perhaps knowledgehood—than any other relation, it is the innermost gatekeeper for category membership. As a consequence, thought about the grounds of membership in a secondary category is always, implicitly at least, thought about how certain things are, in the primary sense, explained. (In the case of a basic natural kind, for example, it is explanatory thought about the origins and maintenance of the causes of the kind's characteristic properties.) When the going gets hard—when the specimens to be categorized are complex edge cases—the explanatory dimension of classificatory inference is liable to become explicit. It begins to matter very much exactly how the relevant properties of a specimen are explained and how their explainers are explained. Further, such considerations are liable to find their way into hypotheses about essential natures, where they are carefully scrutinized and refined. Thus, painstaking exploration of the reason that edge cases are or are not members of a secondary category is painstaking and productive exploration of primary explanatory terrain.

Philosophers ask from time to time whether philosophical investigation is capable of making significant progress (Chalmers 2015; Stoljar 2017).

The views laid out in this book offer some clues to the answer. Yes, substantial progress is possible, but it may well come not in the form of correct theories of the essential natures of philosophical categories, but in a better understanding of the explanatory structures in which the categories make their homes—a better understanding of why waterboarding is wrong, say, or why conforming to epistemic norms leads us to more of the right kind of truth. Just what metaphysical form the structure takes—whether it is all in the end physical, or whether it inhabits swathes of non-naturalistic moral or other normative territory—is a question that I have not attempted to answer. Nor have I given any reason to think that such domains are real, or if real, knowable. Bracketing these foundational questions and skeptical concerns, however, I hope I have given some sense of what philosophical progress looks like.

13.3 The Threat of Emptiness

What if a philosophical concept turns out to be entirely empty, either failing to refer, and therefore lacking a corresponding category altogether, or corresponding to a category that is bereft of members? What is the value of analysis then?

My first piece of advice: do not get paranoid and do not insist on proofs of the impossibility of error. Conceptual emptiness is a normal risk in any kind of inquiry. Science gave us phlogiston and the ether; mathematics gave us naive set theory. Even if these notions were themselves of no ultimate value, they were formulated for good reasons in the course of fruitful inquiry. To have to back out of the occasional intellectual cul-de-sac is a part of the cost of bold thinking.

That said, emptiness can be an astonishing discovery. To find that there is no such thing as causality, knowledge, or the good would surely be an epochal discovery—no? More than a few philosophers have been inspired by the notion that the urge to destroy is also a creative urge.

Against the philosopher-anarchist's credo, consider the following familiar-sounding skeptical argument. Our concepts are constructed to reflect the parochial concerns of the human mind. To discover that such a concept is empty is to discover that something of purely local interest does not exist, a revelation that is hardly exciting.

As it stands, the argument is flawed: the nonexistence of a parochial thing might entail the nonexistence of something of far greater import. It can be given teeth, however, by considering cases where the nonexistence of the parochial thing is entirely due to the world's failure to satisfy some parochially imposed condition. Two examples will illustrate this skeptical strategy.

In late nineteenth-century Europe, it could be argued, determinism was built into the concept of causation. When quantum mechanics suggested with increasing insistence that the world might not be a deterministic place, some thinkers concluded that the world was therefore not a causal place—that there was no such thing as causality.[4] That conclusion is, however, quite compatible with the world being a place where things are stochastically "made to happen." The anti-causality conclusion is less interesting than it looks, then: it reflects a condition imposed on causation in a particular time and place rather than a universal truth about causal metaphysics.[5]

Or consider the notion of knowledge. At some particular juncture in human history, a certainty condition might have been built into the concept of knowledge: nothing not transparently sure to be true can count as knowledge. Skeptical arguments could then expeditiously reveal that there is no such thing as knowledge, or at least that we human beings have little or none. Again, an extraordinary conclusion.[6] But not so extraordinary if it turns out that we have a great deal of extremely well justified true belief. The nonexistence of knowledge in this case has more to do with a parochial insistence on certainty than with the nonexistence of an epistemically far more significant category of mental states.

---

[4] Werner Heisenberg made this argument (Kragh 1999, 209).

[5] A more careful presentation of the argument would have to distinguish two categories: the thing picked out by the late nineteenth-century European concept "causality," and the thing picked out by the concept of causation I am using to entertain the thought expressed in the final few words of this paragraph.

[6] Peter Unger (1979) has made this argument. Perhaps it is no surprise that he later argued that philosophical analysis as it is standardly practiced is incapable of revealing interesting truths (Unger 2014).

13.3. THE THREAT OF EMPTINESS

I do not, of course, deny the philosophical interest of either certainty or determinism. The point of the examples is to show that by analyzing only our parochial concepts, we might arrive at negative conclusions blind to the existence, in the neighborhood, of something of far greater importance. In such cases not only is our conclusion considerably less important than it seems (even if not unimportant); it may create a sense that we are done with the field that leads us to miss out entirely on what matters most. That is the worry that motivates Haslanger and others to advocate conceptual engineering, searching for new concepts that make significant distinctions rather than analyzing old concepts that may perpetuate passing fads or invidious prejudices.

Against the skeptical argument, I renew the comparison between philosophical categories and natural kinds. The examples I gave above turn on ironclad constraints. But if our philosophical concepts are inductive, they contain no such constraints; they consist rather of ordinary beliefs that may be revised in the face of new information about the domain in question.

Thinkers in the late nineteenth century may well have believed that causality was predicated on determinism. If their belief was merely ordinary, however, then upon discovering that the fundamental laws were in part stochastic, it was open to them to reject the need for determinism rather than the existence of causality. That, I propose, is precisely what happened. Just as the Flemish sailors in Western Australia abandoned their belief that swans are white and came to accept that some swans are black, so the physicists and philosophers at work on the interpretation of quantum mechanics came, after a period of reflection and debate, to grant the possibility of probabilistic causality. A similar recognition may be found in the philosophical development of the theory of knowledge: we have relinquished the belief, if we ever had it, that knowledge must be certain.

A parochial belief cannot, then, force us to deduce the nonexistence of a category. It will always be reasonable to ask: might the belief be false? Thus we will always be permitted to jettison the belief rather than the category, freeing our theories to seek out significant explanatory facts however far from our starter theories they may reside. It does not follow that a parochial belief might not under some circumstances bring down a theory inductively. In the light of this possibility I can only repeat my advice to avoid paranoia and demands for infallibility. If a belief is parochial, then there are somewhere out there the resources to see it as such; it is reasonable to hope that, more often than not, it is the prejudices and not the opportunities for profound progress that will, in the end, be spurned.[7]

A final question: does intense scrutiny of edge cases pay off when a concept is empty? That rather depends, I think, on the reasons for emptiness. If a concept's starter beliefs are totally detached from reality, the scrutiny might take you nowhere near any real primary explanatory structure—as would presumably be the outcome of, say, the analysis of astrological terms. But this sort of misfire is, I will say again, a standard risk of any kind of intrepid inquiry. Sometimes analysis will be a waste of time and effort. Often, not. Be happy with that.

13.4 The Concept of Knowledge

It is in the most abstract terms that I have so far talked about philosophical concepts, saying little more than that they are inductive and in some sense explanatory. I now present a worked example: a speculative theory of the concept of knowledge and the related concept of justification. My aim is not so much to correctly describe mental reality (though I hope I am not too far off) as to show how the apparatus that I have developed in this book so far, leaning principally on the psychology of natural kind concepts, might apply in principle to concepts of philosophical categories such as justification and knowledge.

I therefore set out in conjectural mode, more concerned with interesting possibilities than with what little evidence exists concerning the structure of our epistemic concepts. The questions I hope to answer are: What would it mean for the concepts of knowledge and justification to represent those categories as embedded in a network of explanatory relations? And how would analysis of the categories, even if they turn out to be secondary, help to effect the discovery of that explanatory structure? (A third question, as to why the analysis of epistemic categories can be pursued with little or no empirical input, is deferred to the next chapter.)

---

[7] The analysis of categories that belong to false theories is further discussed in Section 14.7. Inferences to a category's nonexistence are treated at length in Section 15.2.

Justificatory Explanation

Odette believes that there is a swan perched on her chaise longue. She is right. Why? How, that is, did she come to hold a correct belief about this matter, as opposed to forming a belief that there was a pigeon on the chaise longue, or that there was nothing on the chaise longue, or refraining from forming any such beliefs?

There is a purely causal explanation of Odette's success. The light was good; her perceptual apparatus accurately generated the visual experience of whiteness, red-beakedness, featheriness, and cygnid shape; she then inferred the presence of a swan from these properties using her accurate beliefs about the characteristic properties of swans.

We can also talk about the case in epistemic as opposed to causal terms. Odette was justified in her perceptually formed beliefs about the properties of the thing on the chaise longue, and she was also justified in her subsequent inferentially formed belief about the category to which that thing belonged—namely, the swans. This description, I suggest, can be transformed into an explanation—an epistemic explanation for the correctness of Odette's belief to parallel the causal explanation above. According to the epistemic explanation, Odette's belief was correct because it was formed in a way that was justified. Most such explanations do not use the term "justification" explicitly. A belief's being justified consists in its standing in certain relations to other things with certain properties; what the explanation spells out are these relations and properties. So to explain Odette's belief's correctness, we might specify its connections to her perceptual beliefs, attending only to such matters as make for justification. This I call a *justificatory explanation.*

A few remarks will clarify the nature of justificatory explanation. First, the thing to be explained in a justificatory explanation is—to make an artificial but expositorily useful distinction—the correctness of the belief, not the truth of the belief. To explain the truth of my belief that, say, ravens are black, is to explain something about ravens; the explanation would presumably proceed in terms of raven physiology and perhaps evolution. To explain why my belief that ravens are black is correctly held is, by contrast, to explain why, in the matter of raven coloration, I apprehend the truth rather than believing something false or remaining agnostic. It is not to explain something about ravens, then, but to explain something about me and my relation to the world.

Second, I have spelled out the form of justificatory explanations for belief correctness without assuming anything about the nature of justification, beyond its proclivity toward doxastic correctness. It might be that justification is closely connected to causation, as on the reliabilist approach. Or it might have nothing to do with processes; a belief's being justified might rather be a matter of its having the right internal logical relations to a special class of "basic beliefs" (such as perceptual beliefs).

Third, it is critical, if justification is to have its explanatory force, that there exists a strong tendency when conditions are right for justified beliefs to be correct. What are these conditions that must be "right" or "not unfavorable" in order for justification to result in correctness? They include a range of things conducive to the success of perception and inference. For visual perception, good lighting may be important. For statistical inference, the evidence must be representative: I am justified in inferring that a tossed coin is fair if fifty tosses yield about one half heads, but I am correct only if the sample reflects the underlying physical probability. For other forms of simple inductive inference, the natural world must be uniform in certain respects. A complete and precise account of the conditions needed for correctness will depend on the nature of justification. Not being an envatted brain is among the required conditions in internalist accounts of justification, but is unnecessary in externalist accounts according to which appropriate connections to the world are part of justification itself.

Finally, there are forms of epistemic explanation other than the justificatory variety—a little more on these presently.

The Explanatory Place of Knowledge

Knowledge plays no role in justificatory explanation. Yet it is closely connected to such explanations, I suggest, by the following principle:

> (K) A belief is knowledge if its correctness is explained by its being justified—that is, if its correctness can be given a valid justificatory explanation.

13.4. THE CONCEPT OF KNOWLEDGE

(Read the "if" like a mathematician, meaning "just in case." But do not presume that (K) is a definition.)

According to (K), a belief that qualifies as knowledge is always justified and true. Gettier's justified true beliefs are not knowledge, however, because their correctness is not explained by their being justified.[8] Consider, for example, Sylvie's belief that Bruno has in his possession a copy of *Twilight of the Idols*. The belief is justified: Sylvie has seen Bruno steal the book. And it is correct: although Bruno thought better of his theft and replaced the book, he has another copy at home. But its correctness is not explained by the justification; the factors that justify the belief, having to do with Sylvie's observation of the theft, have little or nothing to do with the reasons for its accuracy.

A reasoner might treat (K) as spelling out the essential nature of knowledge. But in my inductivist way, I will suppose that it functions psychologically only as an ordinary belief about knowledge, albeit one of broad scope and deep significance.

Principle (K) does not put knowledge at the center of an explanatory starburst, like a basic natural kind, but rather attributes to it a more subsidiary place in epistemic explanatory structure: a known belief is a belief whose correctness is explained in a certain way. Is there some other, more positive explanatory role for knowledge to play?

Arguably there is. Having knowledge explains how we come to have more knowledge. It explains how we come to have more true beliefs. It explains how we come to act successfully in the world. It explains, in particular, the knower's correctly believing certain things. "Why was she able to predict the result of the election? She knows a lot about American politics."[9] These additional forms of epistemic explanation suggest that (K) is only one among several important components of the naive theory of knowledge. At this point, however, I will put knowledge to one side and take a closer look at justification.

---

[8] For a similar view, see work building on Sosa's (2007) virtue epistemology, especially Greco (2010) and Wedgwood (2018). A somewhat different explanationist analysis of knowledge is offered by Jenkins (2006).

[9] Williamson (2000) and Gibbons (2001) offer more substantial discussions of the importance of knowledge as an explainer.

The Nature of Justification

To set out to analyze the nature of justification, and the way in which it explains the correctness of justified belief, is to attempt to lay bare a substantial part of the epistemic domain's primary explanatory structure. What I aim to do in this section is to give some sense of the various possibilities for that structure, by sketching two things that justification might turn out to be. One of the possibilities I wish to present—largely because of its simplicity—is a straightforward vindication of reliabilism. The other possibility, involving the apparatus of entanglement, is something altogether new.

According to the reliabilist approach, a belief is justified just in case it is formed by way of a reliable belief-forming process: valid logical inference, perception by way of well-functioning sensory organs and well-calibrated subpersonal computation, and so on. (I won't attempt any more sophisticated statement of reliabilism than this.) Indeed, its being justified consists in its being so formed; the idemic nature of justification, then, is formation by a reliable process, and to explain a belief's correctness in virtue of its being justified is just to explain its correctness in virtue of its being formed by a reliable process.

The justificatory part of the epistemic domain's primary explanatory structure can therefore be reduced to causal and statistical facts. Talk about justification is no more than talk about certain especially important or relevant physical facts, and justificatory explanation is no more than a certain form of physical explanation especially suited to accounting for the correctness of belief. Or perhaps it is better to say that justificatory explanation provides a kind of framework within which more specific physical stories can be composed, once the details of a belief-forming process and the reasons for its reliability are spelled out. Such a picture is of course the primary goal of naturalistic epistemologists, among whom many reliabilists can be found (Goldman 1986; Kornblith 2002).

An alternative account of justification, also consistent with physicalism, has a rather different aspect. Consider three great rival accounts of justification: the reliabilist account; the traditional foundationalist account, in which justification is having the right logical relation to a certain privileged set of beliefs (the "undefeated basic beliefs"); and the coherentist account,

13.4. THE CONCEPT OF KNOWLEDGE

in which justification is a matter of cohering with the set of beliefs as a whole. All three accounts are compatible with the notion that a belief's being justified can explain its correctness, though they differ on the conditions under which justification leads to correctness. To put it another way, the three accounts of justification identify three ways to explain a belief's correctness.

If the systematic explanation of correctness is a paramount goal of epistemic explanation, then, you might wonder why we have to choose among the three ways. Couldn't each constitute a legitimate variety of primary epistemic explanation? But where, in that case, does justification fit into the picture?

I see several possible answers to this question. First, "justification" might refer to the most important or useful or humanly salient of these three explanatory strategies. Second, "justification" might be an umbrella term referring variously to any of these strategies. In that case, satisfaction of any one of the reliabilist, foundationalist, or coherence criteria would be sufficient for a belief's qualifying as justified; justification, then, would be multiply realizable. Third, "justification" might be a central term in a theory of correct belief formation, represented as entangled with but not necessarily identical to any of the three bases for explaining correct belief. It is this last and most exotic possibility that I would like to explore a little further.

The proposed entanglement can be captured in a single hypothesis having the same surface syntax as a core causal belief:

> (J) Something about justified beliefs endows them with a tendency to correctness.

Core causal beliefs operate under a single-mechanism constraint, as explained in Section 9.6: they zero in on a particular explainer of the characteristic property that they connect to kind membership. With enough knowledge, this explainer—this mechanism—will be represented explicitly, as will the kind's entanglement with the mechanism. Were the single-explainer constraint to be lifted, a core causal belief could be associated with two or more mechanisms, each tending to produce the property in question. As the reasoner learned more about the world, the theory constituting

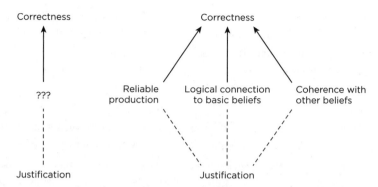

FIGURE 13.2. Left: starter theory of justification, encapsulated in the thesis (J). The dashed line is entanglement; the arrow is a primary explanatory relation. Right: a later version of the theory filled out with multiple entangled properties offering parallel explanations of correctness.

the corresponding concept would come to represent not a single but many entanglements, one for each mechanism.[10]

It seems quite possible that (J), the hypothesis linking justification and correctness, works in this latter way. Without a single-explainer constraint, the property picked out by the theoretical term "justification" in our naive epistemology could be entangled with all three traditional correctness-explainers, reliabilist, foundationalist, and coherentist (Figure 13.2). The term would function, then, as a kind of organizer of correctness explanations, unifying parallel explanations of correctness under a single rubric.

What, in that case, would justification turn out to be? Although it would be (in my current imagining) a secondary category, that does not strictly preclude its having an essential nature. Typically, however, secondary categories such as water and swan have no nature. There are, in other words, no answers to the questions "What is water?" and "What is swanhood?" and likewise, there might be no answer to the philosophical analyst's question "What is justification?" Even so, the category of justified beliefs would be

[10] I am speaking loosely in the main text. Technically, it is unlikely that a substance category will be entangled with two separate mechanisms for a given characteristic property, as that would imply that both mechanisms are normally present. (Consider the case of H_2O and X_yZ, where it is one or the other causing transparency but—unless they are mixed—not both.) In cases like this, the entanglement could only be with a disjunction of the mechanisms rather than with each separately.

real enough, since as a consequence of our inferential dispositions it would have a more or less determinate extension. Thus it would be well able to function as a secondary explainer and as an organizer of primary explanatory structure in the epistemic domain.

The Analysis of Knowledge

Why analyze justification? I gave a generic defense of the value of analysis above; applied to justification it runs as follows: because justification is at the very least an organizer for primary epistemic explanation, its analysis will take us face to face with something of intrinsic interest, the explanatory structure of the epistemic domain. You might think it is rather harder to vindicate the analysis of knowledge. According to principle (K), a belief is knowledge if its correctness can be given a good justificatory explanation. Perhaps that is the essential nature of knowledge, the analyst's ultimate goal. For the sake of the argument, let me suppose so. Would discovering the successful analysis of knowledge, in that case, qualify as a substantial philosophical achievement?

Apparently not. "Knowledge," it would turn out, is the name we give to beliefs whose correctness is explained in certain ways; it is thus an award for explanatory achievement rather than something explanatory itself. What would we have learned? Analysis would have added an item to the epistemic lexicon, but it would have revealed nothing new about the epistemic landscape. Even supposing that this dictionary entry were of some interest, would it have been worth decades of philosophical trench warfare, the slow muddy grind of Gettier and Truetemp, the wasteland of fake barns, the interminable strategizing about opening hours down at the bank, in order to make these precious few feet of progress? There are a lot of fascinating questions out there; was it really so advisable to get hung up on the analysis of "s knows that p"?

Let me conclude the chapter by pushing back against this ennui. One tactic I might adopt is to say: for all we knew when we set out on the analytic path, knowledge would turn out to be an explanatorily interesting category in its own right. Our decision to undertake the analysis of knowledge was therefore reasonable at the time, even if the project turned out to be a disappointment.

That is a decent argument, I think, but I can do better. Consider Gettier. If what I have said above is on the right track, then we discount a Gettierized belief as knowledge because, although it is both justified and correct, its justification does not explain its correctness. As the philosophical literature attests, however, we do not see very clearly why we make the call that we do on Gettier cases. Post-Gettier attempts to analyze knowledge grope toward a definition of knowledge that will make our reasons clear: a Gettierized belief is "lucky" or "insensitive to the truth" or "unsafe" or "based on a false lemma." Or, on closer examination, it is not justified after all. These investigations bring the analyst face to face with a series of questions that lie at the heart of justificatory explanation: How is the correctness of a belief to be explained? What are the different kinds of explanation for belief correctness? And how does justification figure in these explanations? An analyst of knowledge does not set out with the goal of answering such questions, but they find themself making the attempt nevertheless, and in so doing learning about the primary explanatory structure of the epistemic domain.

As with epistemology, so across the philosophical world. The everyday philosophical categories, most of which seem to have been around for at least as long as written philosophy,[11] may be rather far, in the way they carve up their philosophical subject matter, from a fully informed, theoretically sophisticated dissection of the underlying explanatory tissue. We should not treat that as reason to abstain from their analysis. Like the basic natural kinds, the philosophical categories by their inductive-explanatory nature reach forward and down to what is really going on. Even if they themselves turn out to be mere way stations for philosophical inquiry, by examining their explanatory foundations, as we inevitably do in attempting to find their essential natures, unraveling the rationales for category membership in diverse and often knotty cases, we are quite likely, perhaps overwhelmingly likely, to get to the philosophical core of things, to the web of primary explanatory structure. There is no reason to hold back, and—good for us—we have dived right into the upholstery.

---

[11] Machery et al. (2017b) make a tentative case for a "species-typical core folk epistemology" (p. 656). A naive theory of this sort could very well constitute a universal human "starter theory" for epistemic categories such as justification and knowledge (though Machery et al. are somewhat dubious about the universality of the concept of knowledge itself).

CHAPTER 14

Learning without the Senses

14.1 Empirical Indifference

My investigation of philosophical inquiry has been initiated, inspired, goaded, guided, cajoled, urged onward, steered, spurred, and occasionally reined in by the supposition—the conceit—that philosophical concepts are much like concepts of chemical substances and folk genera, and that philosophical categories are much like the kinds themselves: objective, discoverable, and sufficiently deeply embedded into the explanatory landscape of the universe to be worth exploring.

But there is one respect in which many philosophical categories do not resemble the basic natural kinds, a dissimilarity plushly manifest in the metonym of the philosophical armchair. Inquiry into the nature of water is nurtured by, and is surely impossible without, empirical investigation, whereas inquiry into the nature of knowledge—so most epistemologists evince in both words and intellectual action—requires barely even the tips of the philosopher's toes to be dipped into the ocean of empirical fact.

The great gulf between knowledge and water suggests a simple dichotomy; as I wrote in my opening chapter, however, we would be better off regarding the dichotomy's two poles as end points on a spectrum of sensitivity to observable fact, which even at its extremes mixes empirical inquiry and reflection. At one of these extremes are categories such as the basic natural kinds. To even get started on the question "What is water?" you'll want to make regular visits to the chemistry lab. At the same time, chemical knowledge is not enough. There is philosophical analysis to be done, sifting through such puzzles as the distinction between coffee and coffee-tainted dishwater using the method of cases. Pure thought, it seems, is what supplies us with the conditional that is needed to complete the inquiry, that is, the conditional saying that *if* the world has such and such a chemistry (as science tells us it does), then water has such and such an essential nature, or perhaps—if my Chapter 11 argument is on the right track—no essential nature but such and such necessary properties.

At the spectrum's other extreme are philosophical categories whose nature is typically investigated either entirely or at least to a great degree in the armchair: the nature of knowledge, causality, number, the good, and the sublime. A few naturalistic philosophers notwithstanding, most of our colleagues working on these topics find it unnecessary to leave their departmental premises for professional reasons at any time. True, they may unwittingly or deliberately bring empirical knowledge to bear in setting up their inquiry—I have suggested, for example, that the epistemologists call on their empirical knowledge of belief/desire psychology. But the larger and more substantial part of the inquiry then proceeds without any further interruption from the outside world.

Between the two ends of the spectrum lie those many philosophical investigations substantially nurtured by both empirical knowledge and armchair analysis. They include philosophical inquiry into the foundations of the sciences, such as quantum mechanics, statistical physics, and evolutionary biology; philosophical inquiry into metaphysical questions upon which these foundational issues bear, such as the nature of space, time, and physical probability; and in the opinion of many accomplished philosophers, various subdivisions of epistemology, philosophy of mind, and philosophy of language—plus, of course, the kind of philosophy that calls on empirical psychology to illuminate the process of armchair analysis itself.

The treatment of substantiality in the previous two chapters was directed toward categories that lie near the empirically sensitive end of the spectrum, that is, near the basic natural kinds. Leaping over the intermediate cases—I leave methodological questions about the middle ground as an exercise to the reader—I will, in this chapter, tackle head-on the kinds of inquiry at the opposite, empirically insensitive end, examining armchair philosophy in its purest form. Although what I have to say about it may not apply to the rest of the spectrum, the armchair extreme merits a chapter of its own for two reasons: it counts among its instances some of the central philosophical investigations of the modern analytic era, and it poses two inherently interesting, related problems.

The first problem is to explain why the analysis of a category such as knowledge should appear to be for the most part indifferent to empirical information about the mind and its world. I would like to explain, in particular, why some case judgments are accompanied by the sanguine feeling of assurance that I call case certainty (remembering that "certainty" should not be taken absolutely literally). When I count the liquid in my drinking glass as a specimen of water, my confidence is no mystery: the judgment is based on considerable chemical knowledge (my own and others'). By contrast, my conviction that an ordinary Gettierized belief is not knowledge, or that a cause's having a backup in no way undermines its causal status, is apparently made quite independently of my physical and psychological knowledge—and thus is held with a degree of confidence that is undiminished by the recognition that in both domains, I still have much to learn. Why should philosophical concepts function so differently in the cognitive economy than basic natural kind concepts? Might it be that they are not, my best hopes to the contrary, inductive after all?

The second problem is that of substance. Even if philosophical concepts are, like natural kind concepts, in the technical sense inductive—even if at their core they are nothing more than a cluster of ordinary explanatory beliefs—their resistance to empirical correction pushes against the previous chapter's arguments for the substantiality of philosophical knowledge. In their case certainty and more generally their empirical indifference, the philosophical concepts seem more "closed" than "open," more like a stamp imposed on the world than like molten wax taking its shape from the world. Insofar as that stamp is arbitrary, fallacious in its presuppositions, or parochial, so it would seem must be the conclusions we draw from philosophical analysis.

In what follows, I consider two ways to explain case certainty and allied phenomena. The first "topical" approach points to the special subject matter of philosophical knowledge; the second "functionalist" approach points to the special explanatory role of philosophical categories. Although I leave open the possibility that each approach is right for a certain subset of the categories, I favor the functionalist approach; it will, consequently, receive much the larger share of my attention.

14.2 The Topical Explanation of Empirical Indifference

Some philosophical categories—those, above all, with a normative tenor—concern matters that are apparently quite removed from the material world. Consider, for example, case judgments in moral philosophy—judgments about right and wrong actions when it comes to lifeboats, runaway trolleys, intubated violinists, would-be murderers in need of directions, and so on. Once the properties of such a scenario are stipulated, our moral judgment hinges on the reasoned application of certain principles (or so it is usually supposed). These principles, which operate as conditionals connecting material facts about trolleys and tubes to moral valences, enunciate facts about the moral domain alone. Not physics, nor biology, nor even psychology can cast light on the moral domain—so the topical explanation goes. Empirical investigation, then, can have no impact on our judgments about such cases.

From there it is a short step to a "weak" explanation of case certainty, as follows. We are confident in our moral principles (for whatever reason), and therefore confident in the case judgments we make on the grounds of those principles. Because of their special subject matter, neither everyday observation nor scientific research will correct the principles; thus, we expect our confidence to remain unshaken in the light of empirical information. That and nothing more is the nature of our "certainty."

The topical story requires two qualifications. First, though the moral case judgments may be empirically indifferent, moral philosophy as a whole may not be. Aspects of human psychology might bear on moral theorizing if, for example, "ought implies can." Even aspects of evolutionary biology might play a part in the meta-ethical melee (Joyce 2006; Street 2006). That is fine with me; I seek to explain the empirical indifference of moral philosophy, or indeed any kind of philosophy, only insofar as it is actually empirically indifferent. In the present context, I will take the explanation of case certainty to be achievement enough.

Second, the human sciences, and even casual observations of human behavior, provide a kind of information that many philosophers have thought to bear directly on the content of the moral principles, namely the moral views of the world at large. Aristotle, for example, proposed that ethical inquiry must attend to the *endoxa*, usually translated as "currently accepted opinions." That, too, is fine with me, for the same reasons.

I might add, though, that the *endoxa* are a special kind of empirical information: they are information about the end products of other people's armchair reflection. Consulting the *endoxa* therefore does not, in the most important methodological sense, make moral philosophy any less of an armchair pursuit. Likewise, those open to the *endoxa* may experience something akin to case certainty: once all relevant moral opinion is taken into account, they may feel, there can be little doubt about subsequent moral judgments. A 4-year-old has too little experience of their social milieu to be a reliable moral judge, but a thoughtful, well-enculturated adult cannot go wrong. It is to account for this confidence that the topical explanation is devised.

Does it work? Let me take a closer look at its major premise. Ordinary people who experience something like case certainty are reasoning, according to the topical explanation, as follows. My moral principles are claims about the moral domain. Facts about the material domain have no bearing (with the two qualifications made immediately above) on the moral domain. Therefore my principles will not change in the course of empirical inquiry.

There appears to be a strong philosophical presupposition at work in this line of thought: that the moral and the material domain are separate, or if you like, that the truthmakers for moral claims are quite distinct from the material facts. To explain the psychological phenomenon of case certainty, we do not need such a view to be correct; we only need it to be widely held, and then only implicitly. Perhaps it is—perhaps it is a part of our "folk ethics." But I am not so sure; as I said earlier in dismissing hypothetical analysis and the "myth of depth," I doubt that our naive beliefs have any substantial metaphysical content at all.

Further, the topical explanation leaves it rather hard to understand how a naive reasoner—a beginning philosophy student, perhaps—could make a smooth transition to constructivist ethical antirealism or some other view that denies the separability of the moral and the material domain. Their intellectual progress would seem to be toward a view that is in flat contradiction to the implicit folk-ethical grounds of their reasoning. Humans are no doubt capable of such logical acrobatics from time to time, but in the light of my ambition not merely to explain but to vindicate armchair philosophy, the maneuvers look rather dangerous.

As a consequence, I propose not to develop the topical explanation any further. That is not because I see no hope for it. I believe that the topical framework has the resources to provide interesting responses to the questions I have raised so far, and to wrestle with the big question of the substantiality of armchair philosophical inquiry (provided that it can be shown we have some kind of access to the special domains of facts to which the topical explanation closes the empirical doors). In these pages, however, I want to pursue a different approach—the functionalist approach.

14.3 The Functionalist Explanation of Empirical Indifference

The functionalist explanation of empirical indifference draws a parallel between the philosophical categories and what might be called functional categories in the empirical sciences: categories such as force, fitness, and belief. The theories that give these categories their cognitive significance and their extensions have two properties that distinguish them from theories of basic natural kinds, such as our theories of water or swans. First, they have a certain holistic aspect, to be described further below. Second, the causal hypotheses of which they are constituted do not attempt to pick out and describe the effects of particular causal mechanisms; consequently they characterize explanatory roles that can be instantiated by several or many different causal mechanisms, making for categories that are, from a primary explanatory point of view, multiply realizable.

Theories of this sort are not immune to empirical refutation. But as I will show, their response to evidence is an all or nothing affair: they either resist change completely or they are overthrown in their entirety. Further, functional categories generally do not survive their theories' demise. Thus such a category, if it is real or instantiated at all, will have more or less exactly the general explanatory properties that the theory presently attributes to it, though many potentially relevant observable facts remain unknown.

That is why you can investigate the basic structure of functional categories without venturing out of doors.

To illustrate these claims, let me describe two theories that I take to contain functional categories: belief/desire or "folk" psychology and naive or "folk" chemistry. I will then turn to the philosophical categories themselves, focusing on the explanation of empirical indifference while deferring the question of substance to later in the chapter.

14.4 Belief/Desire Psychology

The essential nature of a desire, in the conventional analytic functionalist story, is to have a certain causal role. A part of what it is for a mental state to be the standing desire to show off your marksmanship, for example, is for that state to tend to cause you to shoot any passing swan.

Like many philosophers, I think that the "functionalist" is closer to the truth than the "analytic." My way of honoring this aperçu is to attribute to regular people concepts of belief and desire that are built around functional hypotheses, but to understand the hypotheses as ordinary beliefs rather than as stipulations. According to humans' naive theory of desire, then—according, that is, to folk psychology—the desire to show off marksmanship tends to cause, among other things, swan potshots, but this commitment is hypothetical rather than stipulative, less like a definition and more like the core causal hypotheses that make up, say, the water concept, such as the hypothesis that something about waterhood causes conductivity.[1]

There is one significant dissimilarity. As explained in Section 9.6, the hypothesis about water fastens, in virtue of its underlying semantics, onto a contextually salient mechanism. Here on Earth that will be the mechanism M by which H_2O together with the salts dissolved in naturally occurring water enable conductivity. When the hypothesis's propositional content is

---

[1] Though it plays no role in the following discussion, I should mention that like water, desires do their explaining secondarily: the property of being the desire to exhibit marksmanship does not directly cause, but is rather entangled with the direct causes of, swan potshots. This is how mental states individuated by wide content qualify as explanatorily relevant to the outcomes of narrow psychological processes (Strevens 2008a, §12.2).

made explicit, then, it says roughly *By way of mechanism M, something about water causes conductivity*. The hypothesis is therefore not instantiated by M-less substances such as Putnam's X_yZ, and thus the conductivity and so on of such substances does not incline a fully informed observer to classify them as water.

In the functionalist spirit, I propose that causal hypotheses about desires and other propositional attitudes are not mechanism-specific in this way—they are, you might say, mechanism-generic. (I made the same suggestion about naive epistemology's principle (J) in Section 13.4.) The marksmanship/potshot hypothesis therefore says roughly *By way of some mechanism or other, the desire to exhibit marksmanship causes swan potshots*, allowing that the mechanism may differ from case to case. Such a hypothesis may equally well be instantiated by the cognitive states of a human and of an automaton—call it the PARSIFAL 360—whose thinking is implemented with cogs and pulleys; in both cases, then, a state with the relevant causal profile will tend to be counted by a fully informed observer as the desire to exhibit marksmanship.

Another respect in which belief/desire psychology differs from our naive theories of water, gold, and other basic natural kinds is in its being holistic in two familiar ways. The first kind of holism consists in the fact that any causal hypothesis connecting desire to observable action has among its conditions of application the presence and absence of many other beliefs and desires. A certain constellation of auxiliary mental states must be in place, for example, in order for the desire to exhibit marksmanship to give rise to a swan potshot: the belief that there is a swan in the vicinity, the belief that others consider swans hard to pick off on the fly, the absence of a fondness for swans, and so on.

The second and more important kind of holism consists in the fact that hypotheses about particular desires are typically not assembled from empirical evidence but are rather created by applying a high-level template that has the form of a hypothesis about desire in general. We hypothesize that the desire to exhibit marksmanship creates a tendency to take swan potshots, for example, not because of our (or anyone's) careful observation of aspiring sharpshooters, but because it is an instance of a general hypothesis to which we are strongly committed: *The desire for X tends to cause behavior that increases the probability that X obtains* (here omitting a few obvious if difficult to enumerate qualifications).

As a consequence of belief/desire psychology's holism and multiple realizability, the empirical facts, if they have an impact on desire hypotheses and the like, do so not by tweaking the content of individual hypotheses but by discrediting the theory as a whole, high-level templates and all.

To see this, contrast belief/desire psychology with theories of basic natural kinds, where local amendments—revising a core causal belief here, acquiring one there—are the normal consequences of inquiry. When investigating water, say, empirical facts are relevant in two ways: they help to determine the characteristic properties of water, and they help to determine the mechanisms that cause those characteristic properties. In the case of desires, empirical considerations do not bear directly on either question. There are no facts to be discovered about "the mechanism" by which a desire's causal role is implemented, because the role may be implemented by any mechanism with the right functional profile. There are facts about the analog of characteristic properties for desire, that is, what kinds of things are explained by or explain desires. Some explanatory hypotheses about a given desire are right, then, and some are wrong. But it is difficult to bring empirical evidence to bear on these hypotheses because of the double holism of belief/desire psychology.

So you think that Leda has the desire to exhibit marksmanship but no tendency to take swan potshots? That might be evidence against the hypothesis linking the desire and the action, but it might also be evidence against your attribution of the desire. To show that it is the former, you have three obstacles to surmount. First, you must show that all the conditions of application inherent in the hypothesis hold, that is, that Leda has the constellation of beliefs and desires necessary for the desire to instill the tendency and for the tendency to make itself manifest. That is often difficult, but not impossible.

Second, because the hypothesis is mechanism-generic, there is no mechanism or microstructure that unerringly indicates the presence of the desire.[2]

---

[2] What if there is a mechanism that is characteristic of desire in, say, humans? We could use the absence of the mechanism as evidence for the absence of a desire in any particular case. But in fact we will tend to use it as evidence against the ubiquity, even within the human species, of the mechanism. Compare the case of pain. Were we to find that a certain physical correlate of pain is present in almost all humans, but also to find that a few humans exhibit all the psychology of

Third, and most daunting: if there is something wrong with the marksmanship/potshot hypothesis, then there is something wrong with the template from which that hypothesis was derived—something wrong, that is, with one of your most general hypotheses about desire. It would have to be amended, at the very least, so as not to apply to desires to exhibit marksmanship. Only a huge amount of evidence could convince you to make such a change—so much, I think, that you would find yourself rethinking belief/desire psychology as a whole. The systematicity of the theory, then, gives it a kind of all-or-nothing character as a consequence of which it either stands firm or suffers a catastrophic collapse.

Plausibly, if belief/desire psychology goes, then beliefs and desires go with it: these categories will have turned out to be flawed or empty in the same way that phlogiston, caloric fluid, and ether are flawed or empty. Consequently, the templates describing the high-level explanatory roles of belief and desire are as long-lived as belief and desire themselves: unless we turn our backs on the propositional attitudes altogether, the explanatory templates we now impose on belief/desire psychology will persist in all important respects. For practical purposes they have the robustness, in the face of empirical evidence, of definitions.

Let me emphasize that what I say is true only of hypotheses derived from an explanatory template and therefore concerning the consequences of desires in general—about the effects of, say, the marksmanship desire in any being or system capable of instantiating it. Empirical evidence has a great impact, needless to say, on hypotheses attributing mental states to specific individuals ("Odette would never hurt a swan"), groups ("Americans love their guns"), or even species ("All men desire to kill their fathers").[3]

Holistic robustness gives us what we need for a weak explanation of the case certainty of certain folk-psychological judgments. Suppose you

---

pain without such a mechanism, we would be far more likely to reject the human universality of the mechanism than to deny that the few are in pain. The holistic explanatory structure of functionalist theories overpowers the yearning for simple reductive truths.

[3] As a matter of logical principle there might even be unrestricted generalizations about desire that do not derive from high-level templates and so which lie outside belief/desire psychology's holistic explanatory structure. There is nothing logically incoherent about hypothesizing that all rational creatures desire to kill their parents (if any); its defect lies solely in its falsehood.

are presented with the case of the PARSIFAL 360, and asked whether the machine has a genuine desire to exhibit its marksmanship. (As is typical in a philosophical thought experiment, the facts about the workings of the PARSIFAL 360 are stipulated, so not in doubt; the question is whether something like *that* would have desires.) You deliver a positive answer: yes, the PARSIFAL 360 has such a desire, among many others. And you feel that there is nothing you could learn from neuroscience or cognitive psychology that would change your mind.

The explanation of your empirical indifference has two parts. First, you are confident in the judgment, because you are confident in the belief/desire psychology that you used to reach your conclusion. Second, you apprehend the ramifications of the functionalist story told above sufficiently clearly to believe that the judgment cannot be rebutted by new empirical information. It follows that your confidence in the judgment is for all practical purposes empirically unshakable.

There is a sense in which you lack certainty in your judgment, and indeed, in which it is vulnerable to empirical correction. Were southern Californian neuroscientists to debunk belief/desire psychology itself, you would have to retract your claim, conceding that neither the PARSIFAL 360 nor any other thinker could entertain desires, because there are no such things.[4] Such a qualification is quite consistent, I think, with the feeling of empirical indifference that accompanies folk-psychological case judgments.

As with individual case judgments, so with any investigation based on those judgments: the philosophical analysis of belief and desire will be accomplished largely by reflection, and with little concern for the findings of cognitive science—save for the titanic qualification that if belief/desire psychology as a whole looks to be overthrown, then analysts must certainly rise from their armchairs to pay heed.

---

[4] In fact, the most that the neuroscientists could do, I think, is to show that belief/desire psychology does not correctly characterize the mental lives of humans. If belief/desire psychology were refuted across the board, it would be by theoretical computer scientists showing that the inferential economy of belief and desire is unable in logical principle to drive the sophisticated behaviors that it is supposed to explain.

14.5 Naive Chemistry

The chemical basic natural kinds, such as water and gold, constitute a central part of the categorical organization imposed on nature by our naive chemistry. ("Naive" theories, I remind you, need not be incorrect or even naive; they are simply the theories of a relatively untutored person.) As is the case with their folk-biological counterparts, the chemical kinds are themselves organized into higher-level categories, such as metals and salts. At the very top is the unifying category to which all chemical substances belong—the category of substance itself.

This high-level class is, I suggest, amenable to armchair analysis in much the same way as the categories of belief and desire. We can sit down and ask ourselves: What makes something a substance (in the loosely chemical sense)? What makes a substance different from other fundamental metaphysical categories: an individual, a number, an idea? What do folk substance categories such as gold and water have in common with other "natural kinds" such as folk genera? That each has an essence? A characteristic internal structure? That each is a homeostatic property cluster (Boyd 1988)?

Simply by thinking, we can make progress on these sorts of questions. In doing so, we make full use of category membership judgments that exhibit the same "case certainty" as judgments about Gettier cases or backup causes. Suppose we were to find that chemistry worked quite differently in some other part of the galaxy—there, perhaps, it is Anaxagorean, or Empedoclean. Would something sampled from that sector that possessed the characteristic properties of water be a kind of water? No. But would it be a substance? Yes. No chemical discovery, it seems, could change our minds on this question. Why should that be?

I propose a functionalist explanation. First, the notion of a chemical substance is not mechanism-specific, so something with Anaxagorean or Empedoclean underpinnings is not excluded from the category as a matter of principle.

Second, the high-level notion of a substance provides a template for lower-level theories. The template in question is none other than the starburst structure that causal minimalism attributes to all naive theories of natural kinds, effectively asserting that for any substance, there is a significant set of observable appearances and behaviors to which it is explanatorily related in the starburst fashion.[5]

The notion of substance is therefore ready for almost anything. Individual substances, because conceived of in mechanism-specific terms, may pass in and out of our ontology by the hundreds. But substance itself, because not linked to any particular mechanism, is not so susceptible to empirical eclipse.

A refutation of even the high-level, purely functional part of naive chemistry is, nevertheless, logically possible. We might turn out to live in a world where nothing instantiates the starburst structure.[6] Such a world would be a very strange one: A place where appearances and behaviors cluster consistently merely by coincidence? Or where all stuff is internally homogeneous, but where some outside force orchestrates different appearances and behaviors for different samples? A world of windowless monads? If anything like this were discovered to be the case, we ought ultimately to decide that our philosophical inquiry into the nature of chemical substance is an inquiry into something unreal.

As with desire, then, so with substance: we can find out in the armchair what it is, if it is anything at all. Empirical discovery can tell us that it is nothing, but it cannot amend our beliefs about what kind of thing it is if it is something. That is the basis, again, for a weak explanation of case certainty and of empirical indifference more generally.

14.6 Philosophical Categories

Many philosophical categories, I propose, fall into the same class as desire and substance: they are mechanism-generic and they sit at the center of

---

[5] Further hypotheses about the sorts of appearances and behaviors that tend to be connected in this way might give the general hypothesis more empirical content.

[6] This would constitute a refutation of naive chemistry, I take it, because among its central claims is the assertion that the chemical basic natural kinds such as water and gold are substances. In principle, I suppose, we might reject this assertion and retain naive chemistry as a correct theory of something that does not actually exist. My remarks in the remainder of this section go through, more or less, either way.

sophisticated and systematic explanatory webs, as a consequence of which it is difficult or impossible for empirical facts to impact particular hypotheses piecemeal. In determining their nature, then, the primary role goes to pure thought.

Consider, for example, the principal categories of the epistemic domain, such as knowledge and justification, and the roles that they play in explaining the correctness of beliefs—captured in hypotheses such as Section 13.4's principle (J), according to which something about a belief's being justified confers upon it a tendency to be correct, and principle (K), according to which a belief qualifies as known if its correctness is explained by its being justified.

I take it that the epistemic theory of belief correctness is mechanism-generic, in the sense that it does not concern particular psychological or physical mechanisms for belief formation. (Note that what matters, since we are concerned with the relevance of empirical information, is genericity with respect to causal mechanisms, not with respect to epistemic explanatory structures.) In this respect the epistemic theory closely parallels belief/desire psychology, with which it is of course inextricably entwined: humans and robots can equally well be justified in their beliefs, as can organisms with entirely different sense modalities or faculties of rational intuition—and so the correctness of all such beliefs can be given epistemic explanations, perhaps even the very same explanations.

Further, a hypothesis such as (J) acts much like a template, shaping the structure of an entire class of explanations of belief correctness. It thus forms a part of a systematic explanatory scheme that resists regional empirical revision.

That makes for a weak explanation of case certainty in the same vein as with the other functional categories treated above. Just as we are confident in belief/desire psychology, so we confidently rely on epistemic principles such as (J) and (K). Faced with a Gettier scenario, for example, we reason as follows. Sylvie's justified belief that Bruno has in his possession a copy of *Twilight of the Idols*, true but not for the reasons that Sylvie thinks, fails to satisfy (K): the correctness of the belief is not explained by its being justified. Thus, we readily say that Sylvie's belief is not knowledge.

That is not in itself sufficient for case certainty. What if the theory should change? A new and better principle would replace (K), perhaps pointing to a different decision about the case in question. But it is with the theory of knowledge as it is with functional categories in general: the categories typically do not survive the refutation of their enclosing theories. Our naive epistemology's core explanatory posits, such as (K), will not be amended in response to empirical evidence: either they will remain entirely unchanged or the entire theory, including the categories of knowledge and justification themselves, will be abandoned. There is no way, then, that empirical evidence could show conclusions based on (K) to be false, as opposed to ill-formed or meaningless—as they would be if, say, the very notion of belief turned out to be misconceived.

The ordinary working analyst presumably does not clearly apprehend everything that I have written above about templates and other systematizing explanatory schemes. Yet without understanding the reasons, they can see the consequences: that no easily imaginable scientific finding would undercut the major epistemic principles. The working analyst's conviction as to the empirical immunity of those principles is not as clear, well founded, or justified as a self-conscious analyst's corresponding insight, and they may fail to appreciate in particular that certain empirical findings could bring down the principles in their entirety. But the conviction's force is nevertheless sufficient to explain the sense of assurance with which many philosophers advance their intuitive case judgments.

Is it possible for a judgment about membership of a functional category like knowledge to lack case certainty? Yes, most notably when inductive thinking fails to deliver a sufficiently clear verdict on category membership. I remind you that a functional category's systematizing explanatory scheme—constituted by principles such as (J) or (K)—is not a definition, but rather an ordinary belief about members of the category, albeit one with an especially central role in thought. We make membership judgments not by treating the scheme as a specification of the category's nature, but by using it as a particularly significant piece of information about members of the category and then thinking inductively, using this and other information to decide whether the specimen in question is likely to be a category member. It is entirely possible, in complicated cases, to arrive at an unclear verdict.

That is what is going on when we judge a mental state to be a marginal case of belief, as perhaps when dealing with animals or humans with badly damaged minds. And it is, I propose, what is going on in scenarios such as the marginal Gettier cases presented to ordinary subjects by Turri et al. (2015). Consider this one, for example:

> Emma purchases a diamond from a jewelry store and puts it in her pocket. A skilled jewel thief tries to steal it from her pocket before she leaves the store, but he fails.

Let's say Emma is justified in believing that she has a diamond in her pocket (the chances of an attempted robbery are minimal). Does she know it? The great majority of Turri et al.'s subjects said she does, though many philosophers think otherwise.

Why the disagreement? The question to answer, if principle (K) is to deliver the verdict, is whether Emma's being justified explains her being correct. The issue is not straightforward. Certainly, most of the facts that contribute to her being justified are also a part of the explanation of her being correct: she purchased a diamond from a reputable dealer and put it—as she remembers clearly—in her pocket only a few minutes before. Do they constitute a complete explanation, however? You might ask a similar question about causal explanation: is the thief's failure to steal the diamond a part of the causal story as to why Emma has the diamond in her pocket? Or does the failure render the attempted theft, and so render the failure itself, causally irrelevant to what happened? Such cases of double prevention (or similar) are some of the trickiest in the book.

Knowledge, because it is an explanatory concept, is deeply important, but it is also deeply complex. It ought not to be surprising that on questions such as these, ordinary people and even philosophers are divided.[7]

---

[7] What about the scenario described in Section 10.7, in which the theft succeeds but someone "secretly slips a diamond into Emma's pocket before she leaves the store"? At first it seems that Emma's belief's justification has nothing to do with its correctness, but about half of Turri et al.'s subjects attributed knowledge. Think further about the case, though, and you start to wonder what this surreptitious diamond donor was up to. It seems plausible that they deposited the new diamond only because the old one had been stolen. That gives the scenario more of the character of the failed theft. Further empirical work may show whether or not this explanation of knowledge attribution (when it occurs) is on the right track; what matters here is simply to note the many opportunities for indeterminacy and disagreement offered by complex explanatory questions.

These remarks add to my defense, in Section 10.8, of the thesis that philosophers are better at making case judgments than most ordinary people. There I suggested that philosophers' special expertise is not specifically philosophical but is simply a matter of applying more intelligence and more patience than most reasoners to challenging inductive problems. Now I can add that in many cases those problems turn on questions of explanatory relevance. Thus there can be disagreement; thus there can be lack of certainty. But thus there is also the possibility that certainty, disagreement, and uncertainty may attend the very same case. I'll recapitulate my conclusion that we should not allow contention and doubt about case judgments to undermine our faith in the philosophical analysis of justification and knowledge, any more than we would allow it to undermine our faith in scientific inquiry, which likewise relies on sometimes tricky judgments about explanatory relevance.

By way of a summary, let me conclude this discussion of analytic epistemology by offering a slightly different presentation of the ideas developed above. The deep properties of a functional category will be largely determined by its explanatory schemes. As long as the theory stands, the schemes will not change. Further, the schemes are in the theory and the theory is in the analyst's head. More or less everything needed to answer questions about a category's nature or deep properties, then—though not, contrary to the conceptual analysts' view, the answers themselves—is with the analyst when they take a seat. Provided that the refutation of the enclosing theory remains a distant prospect, analysis therefore may be confidently and conveniently pursued in the armchair.

What explains our ability to investigate the moral domain without recourse to science? A topical explanation was outlined in Section 14.2: the moral facts are not the kind of entities on which empirical information could possibly bear. But as I noted then, to run this line is not as straightforward as it might seem. Let me therefore sketch very roughly a way in which the case of moral philosophy might be eased into the functionalist mold; if I am successful, the topical explanation can be quietly retired.

My starting point, following my treatment of the epistemic categories, is the speculation that our principal moral beliefs map an explanatory network. What is explained? As I suggested in Section 12.2, moral valence itself: why lying is often wrong, why torture is wrong except in a "ticking bomb" scenario, why you should give away a good proportion of your income to alleviate extreme poverty. And also action: why a moderately conscientious individual avoids inflicting misery on others, gives to charity, and lies only about their friends' fashion sense, good looks, and romantic partners. I could articulate a few simple examples of the relevant explanatory principles—causing intense pain to other people is usually bad; ordinary people usually avoid doing morally appalling things—but these are only caricatures of the real thing, and they hardly begin to capture the subtlety of the reasoning of even the most morally obtuse among us.

Different philosophies of the moral domain entertain vastly different conceptions of the way in which moral valence or motivation is explained. But one thing that almost everyone will agree on, I think, is that as with epistemic explanatory relations, the principles do not attach to specific physical mechanisms. Explanations that call on the motivating power of the good, for example, are not restricted to any particular mental substrate: they apply with equal validity to conscientious humans and well-meaning automatons. It does not follow that moral motivation is realized by something nonphysical. That is a possibility, I suppose, but it is equally possible that motivation is multiply realizable, with the motive power supplied sometimes by synapses and sometimes by semiconductors. Likewise, psychological differences might make for moral differences—to put it crudely, what is good for humans might not be good for the PARSIFAL 360—but the relevant psychological differences are themselves multiply realizable.

In addition to its mechanism-genericity, the moral explanatory system is plausibly holistic: it incorporates high-level explanatory schemes relating goods of all sorts to overall moral valence, motivation, and right action. I hardly expect you to be convinced by these vague remarks, of course; for now, I'm happy simply to raise the possibility that moral philosophy's indifference (or near indifference) to empirical information might be functionally explained. The story, then, would follow the functionalist schema: without necessarily having any prior commitments, or even inklings, concerning the foundation of the moral law or the workings of moral motivation, the ordinary reasoner nevertheless senses that, if moral reasoning has any foundation at all, the general principles used to make moral judgments, along with those used to explain moral actions, will resist empirical attrition and so can confidently be applied in ethical philosophy without waiting for the departments of sociology or psychology to issue their final reports.[8]

14.7 Substantiality and Functional Categories

Can philosophical propositions learned in the armchair be reasonably expected to be substantial? In Chapter 1, I sketched the case against. Facts about a category can be discovered by reflection only if the corresponding concept is "closed" with respect to those facts, that is, only if those facts are features of the category that are fixed in advance, never to change. But there is little reason to expect such facts to correspond to aspects of the category that do substantive explanatory work. Indeed, if every one of a category's most central aspects is conceptually closed, there is little reason to expect the category to do any substantive explanatory work at all, or for that matter to be connected in any explanatorily interesting way to the real world. The corresponding concept, I analogized, functions like an industrial die, stamping out a category according to a preconceived plan that, if it happens to match the explanatory or other significant structural boundaries of the universe, does so only fortuitously. To have a reasonable prospect of picking out explanatorily significant features, then, a concept must be "open": it must mold itself to the world, drawing the boundaries of its category in a way that is responsive to the shape of things out there.

Functional concepts such as belief and desire, substance and fitness, knowledge and justification appear to be paradigms of closure. They sit at the center of their theories imposing a certain scheme on the structure of innumerable explanations. The scheme is fixed ahead of time, and will not change: if its presuppositions are found to be false, then almost certainly the

---

[8] Might the functional nature of explanatory moral categories account also for "imaginative resistance," the difficulty in conceiving of scenarios in which the moral facts deviate from actuality (Gendler 2000)?

encompassing theory will collapse and take the functional categories along with it.

It is this fixity that makes the explanatory features of the functional categories available to armchair thinkers; the same fixity, however, reflects rigid presuppositions about the contours of the pertinent explanatory domain. Functional concepts do not mold themselves to the world; they rather take a guess as to what the world is like and then carve things up in accordance with that guess. Such a carving is likely to be not only parochial but factually flawed.

That is the argument against the substantiality of conclusions reached by analyzing functional categories. Let me see if I can subvert it.

Begin with the observation—a crucial one!—that the analysis of functional categories is not truly indifferent to empirical evidence. Functional categories last as long, but only as long, as their theories. Indeed, they are so tied to their theories, and so central to those theories, that the question of the interest of analysis is more or less the same as the question of the interest of the theory itself. The analysis of belief and desire will stand to produce substantial knowledge just in case learning whether or not folk psychology is correct would constitute substantial knowledge.

The same goes for the analysis of justification and knowledge. Would it be interesting to discover that our "folk epistemology" of knowledge, justification, and so on is true? Would it be interesting to discover that it is false? An affirmative answer to both questions implies that the analyses of knowledge and justification themselves are likely to provide substantial knowledge, even if that knowledge (or rather, I should say, epistemic achievement) consists in the revelation that there is no knowledge. However things turn out, we learn something important about the epistemic-explanatory network.[9]

So: folk epistemology, belief/desire psychology, naive chemistry. Interesting if true? Interesting if false? Intuitively I would say that the answers are a reassuring cavalcade of yeses. But let me see whether the skeptical argument from the likely parochiality and unreliability of our starter theories can call this sanguine conclusion into question.

---

[9] Officially, I am taking the substantiality, or at least the interest, of the network itself for granted. Off the record, I am inclined to think that the nonexistence of the network would be a stirring conclusion.

14.7. SUBSTANTIALITY AND FUNCTIONAL CATEGORIES

Our starter theories' explanatory schemes may constitute big gambles on the ways of the world, I have argued, but they are gambles that pay off intellectually even when they lose. The skeptic's best response, I think, is to denigrate the value of such a loss—to contend that it would be relatively uninteresting to learn that our starter theories of belief or knowledge are false, because such a discovery would tell us relatively little about the underlying explanatory network.

The argument follows the line of thought developed in Section 13.3. The explanatory schemes in our starter theories are likely, in their raw state, to contain an unhealthy helping of misconception about the structure of the relevant domain. These misconceptions cannot be corrected by piecemeal revision, because of the schemes' all-or-nothing character. Most likely, then, the enclosing theory will be rejected. Such a rejection tells us that one or more elements of a scheme are mistaken. But that is not very illuminating, because for all we know, the most important elements of the scheme are correct.

Take the case of belief, for example. The starter concept of belief is built around an explanatory template with a number of elements. Were we to discover that all of these elements were gross mismatches with the underlying psychological-explanatory structure, that would be fascinating. The falsehood of belief/desire psychology would not, however, tell us anything nearly so significant; rather, it would merely indicate that something, somewhere in the template is wrong. The all-or-nothing character of the template forestalls a friendly amendment, so the whole thing has to go. But it might well be that a small change is all that is required to put things right—that the correct psychological theory is structured around a very similar template. The obliteration of belief/desire psychology implies that our templates are not *exactly, precisely* on target, but that level of inaccuracy—an eventuality that we ought in any case to have expected—tells us rather little about the workings of thought.[10]

---

[10] A variant on this argument worries that even if the starter template turns out to be true, it might be suboptimal, in the sense that there is a template close by in theoretical space that carves up the explanatory territory in a more natural or more objectively interesting way. There is some additional work in putting the concern in this way: some sense must be given to the idea of a theory's being true yet less good than an equally true rival.

There is something fishy about this argument. Would the falsification of belief/desire psychology really be such a snore? It hardly seems so. The skeptic assumes that the explanatory schemes associated with belief, desire, and other functional kinds are intricate, fiddly things that can go wrong in numerous ways. I disagree. As you can see from my treatment in previous sections, in my view the schemes are in fact rather simple: they assert large-scale, high-level explanatory connections such as a desire's probabilification of what is desired, the starburst structure of substance categories, or the explanatory connection between justification and knowledge.[11]

Why think otherwise? You might reason as follows: numerous attempts to analyze functional categories such as belief and knowledge have yielded complex analyses. If the categories have essential natures, they must be rather baroque. Thus the schemes must surely be rather baroque.

It is an error, however, to conflate natures and explanatory schemes. The schemes are not definitions or hypotheses about essential natures; they do not purport to spell out what it is to be a belief or a piece of knowledge. They do bear on the question of natures, but only indirectly, by guiding the case judgments (and other "intuitions") that are the inductive premises for conclusions about natures. The core causal beliefs about water are relatively simple, yet its nature, if it exists, is almost certainly complex. Likewise, though the explanatory scheme associated with a functional category is uncomplicated, the nature of the category might be positively Byzantine—should that nature exist at all.

Explanatory schemes are, I repeat, broad-based, wide-ranging, high-level sketches of how things work in the relevant domain. They do not commit to specific details (including, as I have said, details of implementation);

---

[11] The conditions of application for such assertions—the ceteris paribus hedges—may in some cases be rather complex, as the case of belief/desire psychology shows. These conditions are not a fussily hand-built supplement to a generalization, however, but are rather—so I argue in Strevens (2012a)—automatically determined in such a way that they cannot go wrong. A generalization that is largely on the right track cannot, therefore, be brought down by a flaw in the conditions of application.

Indeed, ceteris paribus hedges will help to protect hypotheses from the minor predictive and explanatory mishaps that power the skeptical argument. A desire does not probabilify its target if that target is impossible to realize; the right hedge, which in my view is inserted by default, contains this and other necessary qualifications and so ensures that if the probabilification template is wrong, it is seriously and therefore interestingly wrong.

they are free of tortuous footnotes. Further, they connect the category in question to things of undeniable interest, such as motivation and accurate representation. To learn that such a scheme fails to reflect the way things really work is to learn a great deal.

Let me conclude by returning to the transition from the closed nature of a concept to its insubstantiality. A closed concept is like a psychic stencil cutting a predetermined shape from the explanatory substrate, ran the thought, and so is unlikely to reflect the true contours of the substrate. Explanatory schemes are indeed like stencils, with one qualification: it is possible to learn that they are badly shaped. Acquiring this information would be of little import if the schemes were excessively convoluted or randomly assembled. They are not, however: they are plausible, interesting conjectures about the way things work. They might be wrong. But if they are, that would be interesting too.

14.8 How Philosophical Knowledge Is Possible

How can you come to know something substantive when your curtains are drawn, the lights are low, your eyes are closed to the world out there? Here's how: a piece of the world out there is in here, represented by your ordinary beliefs about the philosophical categories.

Such beliefs may be acquired in several rather different ways—biologically, if innate; mimetically, if picked up from parents and peers; empirically, if formed in response to the world. Regardless of their route into the head, however, they can because of their reflexive nature be relied upon to point toward the truth (though the truth may be deflationary or skeptical).

Where the beliefs take you next will depend on their subject matter. If they are like the beliefs that make up concepts of basic natural kinds they might demand, in order to provide theoretical guidance, extensive empirical research. But in other cases—most notably the beliefs concerning functional categories—they enable the pursuit of questions about essential natures and other deep matters from the comfort of the inductive analyst's armchair.

This armchair analysis proceeds by the method of cases. The beliefs issue judgments about cases (sometimes but not always accompanied by the sensation of case certainty), and the case judgments are then used to test theories about essential natures or other philosophical theses. Armchair analysis may proceed hand in hand with empirical inquiry, and even in those cases involving functionalist categories where the armchair alone seems to be adequate to the task, there is typically an empirical hedge hiding behind the arras, waiting if necessary to strike: should the theory in which the functional categories are embedded be refuted, the categories themselves will be cut down and their emptiness exposed.

A quite separate danger also exists, that even where categories are real—even where they have more or less determinate extensions—essential natures, the analyst's ultimate quarry, may not exist. The more common it is for categories to be secondary, the more likely it is that there are no natures to learn.

Yet for all this, armchair knowledge about the categories and the underlying explanatory domain may be efficiently pursued through analysis. Further, for at least two distinct reasons, such knowledge tends to be of substantial matters. On the one hand, inductive theories have a tendency to close in on objective explanatory structure, eliminating falsehood and washing away parochialism. On the other hand, where such a tendency does not exist, as in the case of functional categories, the enclosing theories are a bet that pays off either way: whether the theories are true or false, whether the categories turn out to be explanatorily objective or taxonomical dust in the wind, we learn something well worth knowing.

What, at bottom, supports this edifice of armchair inquiry? Psychologically, our evident willingness to rely on our ordinary beliefs as starting points for every kind of thought. Epistemically—that is a more difficult question. The justification I have offered is built upon the reflexivity of reference, but the fact of reflexivity must be learned from philosophical analysis itself. The best I can do about this circular rationale is to observe that the resulting foundation for analysis is no more precarious than the foundation of inductive thinking in general. At the end of a long day, finding my way home from the psychology lab to the familiar old locus of reflection, that seems quite good enough.

CHAPTER 15

The Life and Death of Secondary Categories

Even were we to acquire a total set of relevant evidence, there are some categories that we would persist in representing (or would come to represent) as chiefly connected to the world by the secondary explanatory relation of entanglement, rather than by primary explanatory relations such as causation or by some overriding nonexplanatory relation such as a definition. These ultimately secondary explainers I have called the *secondary categories*.

The great majority of our high-level categories, such as those of animals, stuffs, persons, and actions, are secondary—so I surmise. Secondary being is the norm at every level other than the fundamental physical level.

There is a great deal that is philosophically and psychologically illuminating to say about these pervasive organizers of thought and substance; I will conclude the book by making a start on that project. I will not have anything to add to my previous theses about philosophical methodology; this chapter exists, then, to elaborate the psychological and metaphysical framework for, rather than to develop further, my ideas about the nature of philosophical knowledge.

All secondary categories are picked out by inductive concepts, though some inductive concepts pick out nonsecondary categories. My first two questions concern inductive concepts in general: How are they acquired?

And how do we decide, in some cases, that an inductive concept is empty—that the corresponding category is uninstantiated or that it does not exist, that there is no such thing?

After laying out my answers, I narrow the focus to secondary categories in particular. I explain why secondary categories (and perhaps some other categories picked out by inductive concepts) are conducive to vagueness. I explain why secondary categories of material things are for the most part irreducible to the fundamental physical constituents of the world. And I conclude by proposing that these irreducible secondary categories are "constructed," and thus that the great majority of the categories we use to systematize the world are in a certain sense creations of the mind.

A few preliminary remarks on the idea of a secondary category. To begin with, the notion is itself a little vague. There are almost no concepts that have their cognitive significance solely in virtue of the putative secondary explanatory relations in which they are embedded, if only because there will almost always be nonexplanatory beliefs about categories: "There are no swans in South Africa"; "The bottle in the back of the fridge contains water." What makes a category secondary is that its concept (fully informed) gets the great part of its cognitive significance from its secondary explanatory role, and that it has no primary explanatory role. I see no need to attempt greater precision than that.

It is important to emphasize, however, that whether or not a category is secondary depends on its explanatory role in a fully informed theory. A category might look to be secondary but may later turn out to be primary if the corresponding theory at first attributes to the category a secondary explanatory role but later, when more is known, represents it as a primary explainer. (Such shifts are discussed in Section 15.4.)

Finally, by tying the metaphysics of secondary categories to properties of their corresponding concepts, I am supposing that only a "secondary concept"—only a concept that represents the category as a secondary explainer once all the information is in—could pick out that category. I will go further: it is in the essential nature of secondary categories to be represented in a certain way. Indeed, to be constructed in a certain way . . .

15.1 Acquisition

A logical approach to psychology suggests that it is impossible to acquire inductive concepts. This evidently deadly problem arises regardless of whether the concepts in question are concepts of secondary categories or, indeed, whether the beliefs that give them cognitive significance have explanatory import at all. That is more than enough to motivate a discussion of the acquisition of—the introduction of—inductive concepts, but the solution to the problem, a new theory of concept acquisition, is also of interest because it tells a particular story about the ways in which categories are psychologically "constructed."

Return once again to that Edenic riverbank where a zoological innocent encountered their first swan. Before the brave new bird swam into view, the naïf had no swan concept whatsoever. After the sighting, they have both a swan concept and a nascent theory of swans, which is to say, according to causal minimalism, a bundle of core causal beliefs or hypotheses connecting swanhood to certain putative characteristic properties—*Swans are white, Swans have red beaks, Swans are strong swimmers,* and so on (understanding these formulations as convenient shorthand for *Something about swans causes whiteness, Something about swans causes red beaks,* etc.). How does the naïf get from "before" to "after"? How, in particular, is their mental term "swan"—a new token representing a kind concerning which until now they had not a thought—introduced into their head?

There are, very broadly, two possibilities; what I have called in Strevens (2012c) the "way of mention" and the "way of use."

When concept acquisition travels the way of mention, the term "swan" first appears in the head in quotes: it is mentioned rather than used. The paradigm of introduction by mention is definition. Define a new term "magwit" and the first appearance of that term in your mind (or language) is in the definition itself, in a mental sentence of the form "Let 'magwit' pick out any individual of above average height and below average intelligence." You may then go on to use the term so defined, and thereby given cognitive significance, to make meaningful assertions: "You are a blundering magwit," and the rest.

More generally, the way of mention begins, always, with a metalinguistic act that treats the new term as a mere token and bestows upon it some sort of semantic property—it could be a definition, an inferential role, an extension—in virtue of which it may from then on be used meaningfully in sentences. (If the coining is silent rather than spoken, the relevant linguistic items and sentences belong to the language of thought.)

A definition gives the defined term an essential nature. In so doing, it creates and delivers up to the definer conceptual truths: magwits are tall by their very (defined) nature, and so are tall of conceptual necessity. In grasping the definition (Rawlsian veils aside), the definer acquires beliefs about the corresponding category that cannot rationally be abandoned. Conceptual inductivism allows no such beliefs; thus, if for some class of concepts conceptual inductivism is correct, it is not by definition that new concepts in that class are acquired.

I conjecture that any plausible version of the way of mention will, because it endows a term explicitly with some or other semantic property, create conceptual truths. Why? Such properties must, I think, put a certain constraint on the corresponding category, so unless the meaning of the term is changed by way of a further metalinguistic act, the category picked out by the term is guaranteed to conform to the constraint.

There lies the problem: a self-aware introducer of the term will form a corresponding belief about the category that is irrefutable—an outcome that is incompatible with conceptual inductivism. Thus, inductive concepts cannot be acquired by the way of mention.[1]

New inductive concepts must be introduced, it would seem to follow, by the way of use: it must be the case that they first enter the head, not in quo-

---

[1] A term might be introduced not by a metalinguistic declaration, but by a metalinguistic hypothesis. Hearing the word "magwit" used by my friends, I may start to guess at its meaning: "Perhaps 'magwit' refers to any tall person." The postulation of such a hypothesis will not in itself create conceptual truths. Semantic hypothesizing of this sort makes sense, however, only when I am learning the meanings of words or concepts that already exist somewhere, in someone's language or head. When a term is introduced metalinguistically for the first time in a linguistic or thought community, its introducer knows very well that it has no prior meaning but is rather a semantic blank; the only way forward is to fill in the blank by declaratively conferring meaning, reference, definition, or something similar. These are the "original introductions" that I am trying to understand.

15.1. ACQUISITION

tation marks, but as functional components of mental sentences. Perhaps, then, my earliest mental deployment of "swan" occurs when, confronted with the new birds, I acquire the relevant core causal beliefs, such as "Swans are white." That, however, seems not to be possible. To form the hypothesis "Swans are white" I must already have the swan concept. So the term "swan" cannot make its first appearance in the head by way of such a hypothesis. The way of use is not available to the inductivist. For that matter, it is not available to anyone: it is simply not a coherent mechanism for concept acquisition.

Let me tighten this dilemma for conceptual inductivism by making a few helpful assumptions about the acquisition of natural kind terms such as "swan." Suppose, first, that the mind knows when it sees its first swan that it needs to develop a new folk genus concept. Eager to do so, it mints a fresh mental predicate "swan." Its problem then is to give that term—which is at this stage entirely meaningless—cognitive significance, and in particular to insert into the "belief box" (or perhaps better, the "hypothesis box") representations such as "Swans are white."

The way of mention proceeds in two steps. In the first step, the new predicate is given semantic properties by an explicit declaration. In the second step, the now meaningful predicate is used to formulate the core causal hypotheses. The procedure is easy enough to implement, but it is incompatible with conceptual inductivism, and in particular with causal minimalism, which is (if I am right) the correct account of the nature of natural kind concepts in adult humans. The way of mention, in other words, is a logically possible means of acquisition that is ruled out, for natural kind concepts at least, by the psychological evidence.

The way of use attempts to omit the way of mention's first step. But how? It can use the predicate to assemble something that looks like a core causal belief—something with the mental orthography "Swans are white"—but in spite of the surface resemblance, this is not the hypothesis that swans are white, because it contains nothing that refers to swans. Indeed, it is not a well-formed hypothesis at all—it contains a nonsense word, "swan," where it should contain a meaningful predicate.

If neither the mention nor the use approach to acquisition is feasible, then what? Perhaps inductive concepts are not acquired at all. Perhaps they are innate. That is Fodor's (1981) conclusion, drawn from an argument related to (and inspiring) the argument above.[2]

That we are born with a swan concept ready to spring into action in case we should happen to run across any such birds seems incredible to most people, and to me too. But sophisticated nativists have always allowed for concepts that are not fully present in the mind until they are "triggered" by the appropriate experience, that is, concepts that exist as potentialities rather than actualities (Leibniz 1981). According to such views, the distinction between empiricism (meaning the view that all concepts are learned) and nativism does not inhere in the distinction between presence and absence—between concepts that are in the head at birth and those that are not—but rather in the distinction between learning and triggering, that is, between concepts that are learned and concepts that make their appearance by way of some other process.

Margolis (1998) shows how a triggering view might work in the case of natural kind concepts. He assumes a psychological essentialist theory of natural kind concepts, but his idea applies equally well to the causal minimalist view. What he posits is not innate concepts, but an innate mental device for building such concepts—a device that will perform the magic trick of giving a term such as "swan" cognitive significance and getting the core causal hypotheses into the belief box without having to confer semantic properties upon the term explicitly.

Here is how the device, which I will call the Margolis module, works (making a few modifications to adapt it to minimalism). An unclassifiable specimen—for example, an organism that fits into no known folk genus—is encountered. A fresh mental predicate is minted; say, "swan." Likely candidates for characteristic properties are noted: whiteness of the feathers, redness of the beak, and so on. And then . . . the moment of conceptual truth. Taking the way of mention, the learner would say to themself something like "I define 'swan' to pick out any white-feathered, red-beaked bird" or "Let 'swan' refer to all organisms with the same DNA signature as those birds over there." Taking the way of use, the learner would not define but hypothesize, saying "Maybe something about swans causes white

---

[2] Fodor's argument concerns the learning of preexisting words, and considers only the definitional version of the way of mention.

feathers"—*per impossibile*, since they have no swan concept to represent the kind concerning which the hypothesis makes its claim.

The Margolis module takes a third way. For each putative characteristic property C, it uses the blank predicate "swan" to construct a mental sentence of the form "Something about swans causes C." This sentence is, as I noted above, defective from both a psychological and a semantic point of view: it contains a term, "swan," that has no cognitive significance and no meaning or reference or other semantic significance. The Margolis module presses ahead regardless and inserts the sentence into the learner's belief box, which is to say, it does whatever it takes to ensure that the learner treats the sentence for inferential purposes as though it were a well-formed hypothesis or belief.

What happens next? Cognitive catastrophe? Not at all. Treating "swan" as though it has cognitive significance endows it with cognitive significance. Once the believer gives sentences such as "Something about swans causes C" the inferential role that their syntactic structure suggests—once they give the term "swan," in particular, the inferential role warranted by the logic of core causal beliefs—they will start to reason as though "swan" genuinely does refer to something like swans. They will, for example, infer (ceteris paribus) that white-feathered, red-beaked, aquatic birds fall under the term "swan," though without the metalinguistic detour that my formulation implies. When they learn more about such birds—that they have double-chambered kidneys, for example—they will populate their belief box with new sentences such as "Something about swans causes them to have double-chambered kidneys." On the dispositional approach to reference, this is sufficient to give the term "swan" an extension, namely, the swans themselves. These belief-like sentences are not, it turns out, defective after all: they have all the usual cognitive and semantic properties of beliefs about swans. A kind of cognitive and semantic bootstrapping provides "swan" with the psychology and the semantics that it needs.

I call this insinuation into the belief box of blank predicates wrapped in hypothesis templates *introjection*. It is by introjection, I propose, that new concepts of natural kinds are acquired (Strevens 2012c).

Some remarks. First, the introjective insertion of an incomplete sentence—a sentence containing a cognitively and semantically blank predicate—into the learner's stock of beliefs is not a rational process. It cannot be, because before the sentence makes it into the belief box, it is, because of its incompleteness, not the sort of thing to which the rules of reasoning apply. In this respect, introjection is like perception: looking at an apple while paying attention in the right sort of way puts the sentence "There is an apple in front of me" into the belief box, but this is not, most philosophers of perception would agree, an inference (at least not at the personal level). Building brains in which such a process occurs nevertheless makes sense; in the same way, building brains in which introjection takes place makes sense, given the advantages of the causal minimalist conceptual structure for thinking about natural kinds.

Second, introjection gives natural kind predicates their cognitive significance by giving them an inferential role. (How else could it be?) But it does not follow that this introductory inferential role has some special epistemic or semantic significance. There is no connection, in particular, between introjection and inferential role semantics: the thesis that natural kind concepts are acquired by introjection is consistent with both the affirmation and denial of inferential role semantics. (As you know, I am a denier.)

Third, where there is introjection there must be, as Descartes might have remarked, an introjector. In the case of natural kind concepts, I have suggested that the introjector is an innate mental device, the Margolis module, which is triggered by the observation of an uncategorizable biological or chemical specimen.

At some level, I think, any introjector must be a part of innate cognitive structure. But that is not saying much: even the psychological empiricists, the barons of the blank slate, postulate an innate mechanism or mechanisms for concept acquisition. The pertinent question is: how general is that mechanism? For empiricists such as Locke or Hume, very general: acquisition is implemented by the same mental process in every domain. The Margolis module is, by contrast, rather specialized, acquiring only basic natural kind concepts. A domain-general introjector is quite feasible, however. Strevens (2012c) suggests that introjection accompanies (rather than preceding or following) the postulation of any unobserved cause. More generally, introjection might accompany the postulation of any unobserved explainer, primary or secondary.

There is much psychology to be done in fleshing out this suggestion. Because my principal aim in this chapter is to show that there is no psychological or logical paradox in the acquisition of inductive concepts, I will rest content with the skeletal structure presented above.

But a few words about concepts of philosophical kinds. How might they be acquired, supposing for the sake of the argument that they are not present in the head at birth? How, schematically, does a language learner pick up the concept of knowledge?

Suppose that a young child hears the word "know" used for the first time in the sentence "She knows a lot about the weather" given as an answer to the question "Why is she so good at predicting rain?" Suppose also that the child is familiar with the words in these sentences other than "know"—"weather," "predict," "rain," and so on. They are then in a position to infer that there is a certain relation a person can bear to the weather, picked out by the word "know" (or "knows a lot about"), that explains why that person is on the whole correct about aspects of the weather. As more information comes in, they can infer that this relation is quite general in its truth-explaining power: for almost any subject matter x, having the "knows a lot about" relation with respect to x helps to explain the having of correct beliefs about x.

The child, let me assume, mints a blank predicate "know" in order to start thinking about this relation. The new term might be introduced into their thought by the way of mention: "Let 'know' pick out the relation that, borne to a subject matter, explains the forming of correct new beliefs about the subject matter." But it might also get there—and far more advantageously I think—by introjection, the arational introduction into the belief box of a sentence of the form "Knowing a subject matter explains the forming of correct new beliefs about the subject matter." The new sentence is treated as a substantive explanatory hypothesis; in this way, a new concept, the concept of knowledge, is planted in the child's mind in the same way that the Margolis module, that tireless natural historical gardener, plants concepts of the basic natural kinds in the thirsty mental soil.

This sort of language-triggered introjection is more or less equivalent to the child's simply "copying" the sentences heard into the belief box (where copying amounts to translating the sentences into the language of thought, inserting blank mental terms in the place of unknown words). Introjection by copying is close to, and perhaps identical to, the process that Carey (2009) calls *Quinean bootstrapping*, in which beliefs about the relations between several unknown categories are acquired by copying the syntactic structure of natural language sentences in which words denoting those categories are embedded.[3]

Perhaps children, especially very young children, copy indiscriminately in this way. My hypothesis is a little more cautious, proposing only that copying occurs when the sentences crossing the society/mind barrier concern explanatory relations between the unknown and the known categories. Either way, I surmise, introjection is the principal means by which new explanatory concepts are acquired.

15.2 Discovering Nonexistence

There is no such thing as phlogiston, or the ether. There is no such thing as N-rays. There are no yetis, rocs, or mermaids. How do we reach such conclusions? Why do we not conclude that mermaids are simply the members of the order *Sirenia* (manatees and dugongs)—not quite what the sailors were hoping for, perhaps, but perfectly fleshy and real?

One way to make nonexistence judgments is by thinking explicitly about the semantics of the term in question. Suppose that "mermaid" is defined to mean "half-human, half-fish creature inhabiting the seven seas"; seeing that nothing satisfies this description, you can conclude that mermaids do not exist. Or suppose that you know that "mermaid" was introduced with the intention of picking out those ravishing beings of the same kind as certain baptismal specimens dimly sighted from the masthead after the distribution of the daily ration of rum. The specimens are real enough—they are members of *Sirenia*—but because they fall short by human standards of beauty, a presupposition of the referential intention is false and so the act of reference-fixing falls through. You conclude that "mermaid" has no extension: it is empty.

---

[3] Carey suggests that concepts acquired by Quinean bootstrapping come with meanings of the conceptual role variety. If so, Quinean bootstrapping is not introjection. The psychology of Quinean bootstrapping seems to me, however, to be entirely detachable from this semantic hypothesis; I would therefore favor an amendment to Carey's powerful theories of concept acquisition—of the acquisition of mathematical concepts by children in particular—that retains the bootstrapping in an introjective form.

15.2. DISCOVERING NONEXISTENCE

In the first of these cases, mermaids are a perfectly well-defined natural kind that has no members. In the second, there is no kind picked out by "mermaid." We are all agreed that there are no mermaids, but do we intend this assertion of nonexistence in the first way or in the second? Is the term "mermaid" empty, or does it refer to a kind that is determinate and real but in this world unpopulated? Normal thinkers, I suggest, have no definite opinion on the matter. They could not say whether their theory of mermaids is true but that nothing falls within its scope, or whether the theory is semantically defective and so can be neither true nor false. This suggests that, in reaching the conclusion that mermaids do not exist, normal thinkers travel neither of the semantic paths described above, both of which require the thinker to resolve the question of empty category versus empty term en route. How, then, do they get there? Inductively.

You can infer the nonexistence of mermaids using inductive reasoning in the same way that you infer the nonexistence of anything: you acquire evidence against those theories in your hypothesis space that imply the existence of mermaids. In the case of a putative basic natural kind (let me treat the mermaid scenario as such for the sake of the argument), you have just one theory implying the existence of mermaids, a set of core causal hypotheses and associated beliefs including *Something about mermaids causes them to have human torsos, Something about mermaids causes them to call out to sailors, Something about mermaids causes them to swim near rocky shores*, and so on. If this theory loses credibility, then your belief in the existence of mermaids will lose credibility by a commensurate amount.

It may help to put this more formally, in loosely Bayesian terms. Let t be your theory, or set of beliefs, concerning mermaids. Let m be the posit that mermaids exist. Then your subjective probability for the existence of mermaids $C(m)$ can be broken into two components: $C(m) = C(mt) + C(m\neg t)$. Suppose that the theory entails the existence of mermaids, so that $C(mt)$ is equal to $C(t)$. Suppose also that you have no other reason to think that mermaids exist, nor later acquire any, so that $C(m\neg t)$ starts out and remains minute. Then as the probability of the theory declines, the probability of m will tend to decline likewise: as t appears ever more dubious—as you begin to doubt the existence of the comely faces and the siren songs, and to attribute the whole story to poetic fancy—you will begin to doubt the existence of mermaids themselves.[4]

Yet it does not always work out that way. As the parable of the schwanns shows, we sometimes bootstrap our way out of an old and largely incorrect theory of a kind to a newer, better theory: though the first schwanns to be observed were abnormal, sporting pink feathers and blue beaks, we eventually came to believe that schwanns have green feathers and yellow beaks (Sections 5.3 and 9.1). Why is our theory of mermaids not transmuted, in the same way, into a biologically respectable theory of the *Sirenia*?

It is tempting to give an answer that turns on semantic properties or conceptual structure. Perhaps the human visage is built more deeply into the mermaid concept than the pink feathers were built into the schwann concept? But there is no need for such maneuvers. Epistemic differences—differences in the information we possess and its evidential bearing—are enough to explain why mermaids are headed for conceptual extinction while schwanns, in their new guise, thrive. The paramount difference is that, in the case of the schwanns, there is a group of individuals who are clearly identified as schwanns and who can be more closely examined to test the schwann theory. Had the "mermaid" designation stuck to particular manatees, we might call these creatures mermaids today.[5] (In this connection I am obliged to note that "dugong" may be derived from a Malay expression meaning "lady of the sea.")

Against the inductivist story, the following complaint can be lodged: once the theory of mermaids t is discarded, there remains a small but nonzero probability $C(m\neg t)$ that mermaids exist. But in fact (continues the objector) we infer conclusively that mermaids do not exist: once the theory goes—once we abandon the belief that mermaids are half-human, half-fish, for example—we have no credence in the existence of mermaids whatsoever. The same objection might be made even more forcefully in the case of abandoned theoretical notions such as phlogiston: once the

---

[4] Bayesian aficionados know that there is nothing about the structure of the subjective probability function to guarantee that $C(m\neg t)$ will remain small as $C(t)$ declines. This is rather an additional postulated property of $C(m\neg t)$, plausible but not policed by the basic Bayesian machinery.

[5] Then again, the fanciful nature of that word might lead to its being abandoned in favor of some more sober term. There is a big difference, however, between a term's falling into disuse and its figuring in a nonexistence claim.

phlogiston theory of combustion is relinquished, we conclude just as surely as the theory is false that phlogiston does not exist. It sounds absurd, after all, to speculate that phlogiston might yet exist although everything that phlogiston theory says of it is false.

Such considerations revive the thought that the semantics of "mermaid" and "phlogiston" are somehow bound to their respective theories—that the phlogiston theory of combustion, for example, constitutes a tacit definition or reference-fixing description of "phlogiston."

I have already given you a reason to reject this view: if the existence of phlogiston is denied on the grounds of an argument taking as a premise the semantics of "phlogiston," then a clear-thinking phlogiston-denier should be able to tell us whether the term is empty or merely uninstantiated, having made up their mind in the course of concluding its nonexistence. Even the clearest thinker cannot. But put that objection aside for now; what can the inductivist say about the motivating datum? How can an inductivist explain why we reach nonexistence judgments about mermaids and phlogiston with the level of certainty that we seem to feel?

I will not deny the existence of irrevocable judgments of nonexistence. They are made by at least some thinkers, I believe, and they are to be explained as follows. The phlogiston-denier comes to the conclusion, on the basis of overwhelming empirical evidence, that phlogiston theory is surely false (no problem for inductivism there). They then observe that the only connection between the word "phlogiston" and the world goes by way of the theory. Given the falsehood of the theory, then, there can be no connection—thus, "phlogiston" must fail to pick out any existing thing.

The story can be augmented by supposing that the phlogiston-denier subscribes to the dispositional doctrine of reference, and so believes that anything that would not be counted as a category member given a total set of relevant evidence is not a category member. Without phlogiston theory, the denier will reason, there can be no disposition, even upon arrival of all the relevant evidence, to apply the term "phlogiston" to any actual substance; it follows from the dispositional doctrine that no actual substance falls within its extension.

My explanation of irrevocable denial turns, as you can see, on the denier's explicitly reasoning about semantics, and about reference in particular. This is strong medicine for the inductivist, permitted with a philosophical prescription but to be used only very sparingly. Let me see how little I can get away with swallowing, in two ways. First, I suggest that most normal judgments of nonexistence do not make it as far as the semantic train of thought described above. An ordinary working chemist sees that phlogiston theory is false and so rejects the existence of phlogiston; they do not further reflect on the nature of reference to decide that the rejection is irrevocable.

Second, although the explanation posits reflection on semantic matters, it does not suppose that the thinker is in possession of a theory of the nature of reference. The dispositional doctrine, which is not itself a theory of reference, determines that nothing falls under "phlogiston," but it leaves open whether phlogiston is a real but uninstantiated kind or whether "phlogiston" fails to pick out any kind at all, a question that would presumably be settled by a comprehensive theory of reference. That is just as well since, as I observed above, even sophisticated reasoners and enthusiastic deniers seem a little unclear on such matters.

Why, you might wonder, does my putative phlogiston-denier subscribe to the dispositional doctrine? My best guess is that the dispositional doctrine inheres in our naive semantics, that is, in our folk theory of semantics. I doubt that the folk theory of reference is the dispositional theory, however. An intriguing possibility is that the folk theory is rather the simple descriptivist theory, which as I showed in Section 9.7 implies the dispositional doctrine—a psychological hypothesis that would explain descriptivism's intuitive grip. In that case folk semantics would be, like folk physics, deeply false in its fundamental principles, yet rather useful all the same because of the many valuable truths that it implies. But I advance this idea only very tentatively; more would have to be said, for example, as to how the descriptivist theory accounts for the inability of even the most self-assured deniers to distinguish between empty and nonexistent categories.

15.3 Vagueness

A term or concept is vague if it admits of borderline cases, that is, if there exist or could exist specimens that neither determinately fall under the term nor determinately fail to do so. "Dirty" is a paradigm: there are people in mild states of filth who are not clearly dirty and not clearly not dirty. They inhabit the territory between dirtiness and non-dirtiness over which neither property authoritatively rules.

Conceptual inductivism, and the notion of a secondary category in particular, can shed some light on the psychology of vagueness, that is, the psychological reasons for the appearance of the existence of borderline cases. (In what follows, the semantics of vagueness will be discussed only in passing. The logical problems posed by vague terms, such as the sorites paradox, which are the principal spur to most philosophical work on vagueness, will not be discussed at all.)

Begin with two approaches to the psychology of vagueness, epistemic and metaphysical.

According to the epistemic psychology of vagueness, the appearance of borderline cases is misleading: what seems to be an indeterminacy in category membership is in fact uncertainty about category membership. Scrutinizing the not-quite dirty man, unable to pronounce him either dirty or non-dirty, you might take yourself to be judging that he is neither determinately dirty nor determinately non-dirty, but in fact you are simply unsure whether the man is dirty or not. A vivid realization of the epistemic psychology is the epistemic account of the logic of vagueness. In this view (Williamson 1994), there are no borderline cases. There is always a matter of fact as to whether a person is dirty, for example, though that matter of fact is typically not clear to us. We do not know whether the man before us is dirty, but nature knows.

According to the metaphysical psychology of vagueness, we judge that a specimen is a borderline category member because our concept harbors a criterion purporting to spell out the category's essential nature, and we see that this criterion either makes no determinate judgment concerning the specimen or determinately judges it to be a borderline case.[6]

In the psychology literature, the technical notion of "graded membership" has been used to construct a metaphysical psychology of vagueness (Hampton 2007; see also Murphy 2002, 20–22). Membership in a category such as "dirty people," according to these views, is in intermediate cases not

---

[6] The criterion might be stipulative and therefore apodictic, or it might be a mere essence postulate. The explanation of judgments of vagueness is "metaphysical" either way.

an all or nothing matter, but is captured by a number—running from, say, zero to one. (Perhaps only a few extreme specimens earn zeros and ones.) The claim "x is dirty" is then understood to mean something like "x's membership grade in the category of dirty people is near one," while the claim "x is neither determinately dirty nor non-dirty" means "x's membership grade in the category of dirty people is near one-half." (A sophisticated view would allow for contextual effects.) There are, then, two systems at work in such a psychology of vagueness: the system that assigns membership grades, and the system that maps the fine-grained scheme of graded membership in the mind onto the coarse-grained scheme of all-or-nothing membership in natural language. The criteria that assign membership grades to specimens are typically assumed to be definitive, so that what is judged to be a borderline case (or better, an intermediate case) is indeed such—though the question whether borderline cases are real or only apparent is not psychologists' primary concern.

What does conceptual inductivism have to say? If a concept is inductive, then judgments concerning membership in the corresponding category are typically inductive—often proceeding along the lines of inference to the best explanation. But inductive reasoning leads to conclusions with probability rather than certainty, and in some cases leads to no determinate conclusion. An inductive categorizer might well, then, end up unable to decide whether a specimen fits a category. In such cases, the indefinite opinion about category membership is explained epistemically rather than metaphysically.

Do any interesting or central cases of vagueness arise in this way? Some, at least.

Consider the evolutionary sorites (Section 10.3). Swans evolved from non-swans. Follow the lineage of a present-day swan far enough into the past (say, along the female line), and you find a non-swan ancestor. Now reverse the journey. At some point you go from the last of the non-swans—call them *proto-swans*—to swans. But where? Does it happen in a single generation, with a proto-swan mother laying a swan egg? Or is there a period of transition? If the latter, how do things stand with respect to the swanhood of the progenitors in that intermediate phase?

This is surely a classic case of vagueness: the transitional progenitors are, or appear to us to be, borderline cases of swanhood. Of any one of these links in the chain of swan evolution, we would be reluctant to say either "Yes, it is a swan" or "No, it is a mere proto-swan." Why?

Not, I suggest, because we have in our minds a criterion for swanhood that assigns to these specimens an intermediate grade of category membership. Rather, the psychology is this: we have a theory of swanhood, a set of core causal hypotheses and associated beliefs about swans, from which it is impossible to infer with any confidence either that a transitional specimen is a swan or that it is not. This is the characteristic uncertainty of inductive logic: the thing is sufficiently unlike a swan that we are unconfident that it is a swan, and sufficiently like a swan that we are more or less equally unconfident in ruling out its swanhood.[7]

An epistemic psychology such as this does not entail an epistemic approach to the semantics or metaphysics of vagueness. It is not necessary, that is, for a psychological epistemicist to follow Williamson in postulating a matter of fact about the swanhood of the transitional links. There may exist no matter of fact, yet our immediate doxastic relation to the swanhood of the transitionals may be not so much to grasp the existence of the factual void as simply to find ourselves with insufficient grounds to form the sort of belief that would fill it. Likewise, a psychological epistemicist need not posit that we believe that there is a matter of fact about the transitionals' swanhood, or that we believe more generally that categories have sharp boundaries.[8] Ordinary people may have no opinion about the matter.

---

[7] There is another evolutionary sorites arising from possible rather than actual evolutionary histories. Suppose that the proto-swans had evolved to have yellow rather than red beaks. Would they still have been swans? Surely. What if they had evolved to have yellow rather than white plumage? To avoid large bodies of water? To give birth to live young? To have arms rather than wings? The answers, confidently affirmative near the beginning of the list, are confidently negative near the end. But there is great uncertainty in between.

[8] Bonini et al. (1999), in their presentation of a psychologized version of Williamson's theory of vagueness, do posit such a belief, though it seems to me not to do much explanatory work. For these authors, the uncertainty that generates borderline membership judgments is uncertainty about the precise boundaries of a category, whereas in my approach, it is uncertainty about the membership of some particular specimen. I conjecture that the authors are somewhat under the spell of the "myth of depth" (Section 6.4): informed categorizations are typically based, they believe, on a principle that purports to state the ultimate basis of category membership. Such a principle reveals whether or not a category is inherently vague—that is, whether it admits of borderline cases. Since, according to the myth, competent categorizers have a definite belief about the form of the principle, they have a definite belief about whether or not the category is vague. A psychology of vagueness

That is not the end of the story, however. Williamson's epistemic theory attracts the incredulity it does because philosophers and their friends find it hard to believe that there could be a fact of the matter about dirtiness, or swanhood, in every pertinent case. Something persuades us that for certain specimens, no further information could possibly settle the issue. We conclude that we lack the information to classify a transitional swan progenitor as a swan or a non-swan because there is no such information to be had.

A psychology of vagueness ought to account for our endorsing this conclusion, whether or not we are correct to do so. What convinces us that there is not some further fact out there that inscribes a neat, fine line through the ancestral swan lineages, separating the first swans from their non-swan mothers? I would like to give an inductivist answer to this question that calls on the same resources as my answer, in the previous section, to the question of how we come to deny irrevocably the existence of mermaids and phlogiston—namely, ordinary people's commitment to a naive semantics that includes the dispositional doctrine of reference.

Well-informed philosophizers are able to see that, even if they received a total set of relevant evidence, they would not be able to draw, with confidence, a line precisely separating swans from proto-swans. Subscribing as I hypothesize we all do to the dispositional doctrine of reference, they then conclude that there can be no fact of the matter as to whether or not these empirically undecidable cases fall inside the swan concept's extension. (Without possessing a total set of evidence, they may not be able to say exactly which cases are undecidable, but once they know enough they can see that, however the evidence turns out, there will be undecidable cases.)

From psychology back to philosophy: are categories such as swan really vague, or do we merely suppose that they are? That vagueness is real follows from my preferred formulation of the dispositional doctrine, according to which a specimen is determinately a category member or nonmember only if we would determinately judge, upon receipt of a total set of relevant evidence, that it is a member or nonmember. Other theories of reference,

---

must therefore tell us which way, for ordinary people, that belief goes. This is precisely the kind of metaphysical belief or essence postulate that, according to my inductivist approach, plays little or no role in the psychology of concepts.

15.3. VAGUENESS

even other versions of dispositionalism, might with some effort be framed to give the contrary answer, but I will suppose that vagueness is genuine and pervasive.

The ultimate causes of vagueness are mental theories that, even when fully informed, are unable to decide category membership in certain classes of cases. In the remainder of this section I want to explain why theories of secondary categories are especially prone to fully informed undecidability, and thus why secondary categories are especially prone to vagueness.

The first and most obvious explanation for undecidability is that the decision procedure for secondary category membership is a complex inductive matter. Consider a chemical sorites: a container of concentrated coffee sludge sits above a container of water, leaking. As time goes by, the liquid in the lower container goes from paradigmatically watery to something indistinguishable from coffee. Choose an intermediate time. Is the stuff in the container at that time still water? Or is it coffee (or coffee-flavored beverage)?

This is a question to be settled—even for the empirically fully informed —by inductive logic. Is it waterhood or coffeehood that best explains what you see before you? (As in other cases, intentions make a difference. Is this the latest pricey Japanese coffeemaker? If so, you may be inclined to shift your judgment coffee-ward sooner than otherwise.) For intermediate cases, you will have insufficient grounds to decide the issue either way. When there is no prospect of further relevant information to settle your uncertainty, you declare the specimen to be a borderline case.

Let me give, in the remainder of this section, two further features of secondary categories that contribute, epistemically, to the uncertainty that spawns borderline cases, staying with basic natural kinds for the sake of concreteness.

Water is transparent. That a liquid is transparent, then, gives you some reason to think that it is water. But what counts as transparency? At what point does the increasingly caffeinated container of water go from being transparent to not transparent? Transparency itself is a vague notion, and this vagueness will tend to be passed on to any category whose cognitive significance inheres in its explanatory connection to transparency. Ineliminable vagueness is not inevitable, as would be the case if transparency figured in the category's definition, but it is likely if the category has no idemic nature or has a nature determined by its explanatory role.

Both secondary and primary explainers of vague derived properties, then, will incline to vagueness themselves—secondary explainers especially, since they are less likely to have substantive essential natures that trump explanatory considerations in deciding category membership. This contagion cannot explain how vagueness originally arises, of course, so although worth noting, as a transmitter rather than a generator it is not of the greatest importance in explaining vagueness.

A related phenomenon—a kind of circularity—is a true generator of vagueness. Let me return to the case of the swan progenitors, those transitional forms floating uncertainly between swanhood and the determinate lack thereof. A typical progenitor is a lot like those birds that we regard as unambiguous swans—the mainstream swans—but subtly different. Perhaps its neck is just a little less curved than the necks of mainstream swans, for example. What does that tell us about its swanhood?

The crucial question: is its neck curvature explained in the same way as the curvature in mainstream swans? A necessary condition for sameness of explanation is that the transitional's lesser neck curvature is produced by the same proximal mechanism as mainstream curvature. Call the latter the *swan curvature mechanism*; the question, then, is whether the curve in the transitional bird's neck is produced by the swan curvature mechanism.

If the mechanism in the transitional bird were identical to the mechanism in mainstream birds, the answer would of course be yes. But for the sake of the argument (and so as not to postpone a problem that will sooner or later arise), suppose that although it is similar, it is not exactly the same as the mechanism in any mainstream bird. Does that mean that it is not a realization of the swan curvature mechanism? Not necessarily: that mechanism might be individuated broadly enough that it includes among its possible realizations the transitional mechanism.

In order to decide, we should look to our standards for individuating explanatory mechanisms, that is, to our theory of explanation. Given an explanandum, the theory of explanation will specify an explanation—in this sort of case, a causal mechanism. If the explanandum is swan neck curvature, our theory will specify, as its causal explanation, the swan neck curvature mechanism. The issue, then, is whether that specification is broad enough to include the transitional curvature mechanism.

You might think that our standards for individuating explanations are far too loose to give a definite answer to such questions—that even given a precise explanandum they will put only rather sloppy constraints on the explanatory mechanism. I think this is wrong: our standards are quite precise (Strevens 2008a). But even if you will not believe that, accept it for the time being in order to help me pinpoint a separate source of indeterminacy. Indeed, for that reason accept something that even I would not insist on: given an absolutely precise explanandum, our standards for explanation determine an absolutely precise mechanism specification, that is, a mechanism specification with no slippage whatsoever. Such a specification might be quite broad—it might be realizable by a wide range of mechanisms—but it will have an exact cutoff point, a boundary where you go from explanatory to nonexplanatory with just a minute adjustment in causal structure. If all of this is correct, then there could be a determinate matter of fact as to whether or not the transitional curvature mechanism counts as a realization of the swan curvature mechanism.

But there could not be. The mechanism specification will be precise only if the explanandum is precise, and in the case at hand, the explanandum—swan neck curvature—is not precise, for a very interesting reason. Exactly which range of geometries that physiological property spans depends on which animals are swans. If the transitional birds count as swans, then swan neck curvature includes the lesser curvature of those animals' necks. If not, not. But whether the transitionals are swans is the very question that was at issue. This is the circularity that makes it impossible to nail down determinately any of the following: what counts as swan neck curvature; what counts as the mechanism that explains swan neck curvature; what counts as a swan.

It would be fatal, this circularity, if the notions in question—swan, swan neck curvature, swan neck curvature mechanism, and so on—were defined in terms of one another, as the definitions would not be well grounded. But because some of the connections—in particular, the connections between swanhood and the rest—are explanatory rather than defining, you have nothing more than the kind of local holism that typically accompanies inductive inquiry.

In fact, this circularity, far from being objectionable, lends the inductive enterprise a certain beneficial flexibility, in virtue of which our hypotheses about neighboring theoretical categories are continually mutually adjusted in the light of new information without the need for some semantic legislator to step in over and over again to redefine the vocabulary. It is a wonderful flexibility that any wise investigator would want to incorporate into their cognitive structure. The price is pervasive vagueness.

15.4 The Irreducibility of Almost Everything

The material universe is built, evidently, from elementary particles (or quantum fields or strings) and their spatiotemporal, causal, and other nomological interrelations. You might expect, then, that every aspect of material being—every material property, category, or entity—would be reducible to these fundamental things. Or to choose a particular characterization of reducibility, you might expect that what it is to fall into any given high-level category or to instantiate any given high-level property could be specified in a language referring only to fundamental-level entities and relations.[9] But many, perhaps most philosophers concerned with such matters believe it isn't so. Everything material is constituted of fundamental-level stuff, yet almost nothing can be reduced to that same stuff. How can it be?[10]

Let me sketch an explanation of pervasive irreducibility in the material domain, focused for expository simplicity on high-level categories of an explanatory nature, that runs as follows:

1. Most concepts are inductive.
2. Most inductive concepts start out as secondary concepts: they are introduced into the mind embedded in secondary explanatory relations.

---

[9] The language must of course contain the usual logical connectives and the sort of technical apparatus that, for example, implements the weighting schemes used to formulate category membership criteria by the prototype approach to concepts.

[10] If what I say here is correct, irreducibility may be the rule in other domains too—it may be that high-level moral categories, for example, are mostly irreducible to the morally fundamental level. My focus throughout this section is, however, the material domain.

15.4. THE IRREDUCIBILITY OF ALMOST EVERYTHING 317

3. Once embedded in secondary relations, in a universe like ours a material concept will tend to stay that way even on receipt of a total set of evidence. Thus most inductive, material category concepts pick out secondary categories.
4. Most secondary categories are irreducible to the fundamental level.

Putting it all together, and not pausing overlong to worry about the cumulative effect of concatenating a string of "most"s, it follows that the great majority of categories of material things are irreducible to the fundamental physical level.

I have four premises to justify, or at least to persuasively assert. The first is that most high-level category concepts are inductive. That follows from my generalized conceptual inductivism; I won't say anything more in its support.

The second premise states that inductive concepts tend to start their mental lives embedded in theories that represent them as connected to the world largely by secondary explanatory relations—that is, by entanglement with causes and other primary explainers. This is true, I conjecture, because inductive concepts are acquired by introjection, and it is characteristic of introjection to take the cautious strategy of connecting new categories, at first, by way of secondary rather than primary explanatory connections to the properties that they are supposed to explain and by which they are supposed to be explained. The acquisition of basic natural kind concepts by the introjection of core causal beliefs is a paradigm of this conservative approach.

Perhaps not every inductive concept is acquired by introjection; perhaps some—the tools of "core cognition"—are present in the mind from the start. For much the same reason that introjection is cautious, I conjecture that many of these inborn representations are also cautious, opting to hypothesize secondary over primary connections by default.

My third premise is that a material category concept (a concept of a category of material things) that enters the mind embedded in secondary relations will tend to stay that way—that even as the reasoner learns more about the physical world, even as their enclosing theory becomes larger, more sophisticated, deeper, the represented connection between category and theory tends to remain one of entanglement. Consequently, a concept that starts out looking "secondary" will in most cases remain secondary even on receipt of a total set of evidence, and so will pick out a secondary category.

In order to account for this tendency, let me proceed indirectly. I will ask: By what process can an initially secondary explainer come to be thought of as a primary explainer? What do we need to learn about a category to upgrade its explanatory status, to discover that it is doing the primary explanatory work—the causation, the justification, or whatever—itself, rather than through some entangled proxy? I then make a case that such upgrades are rather rare.

Suppose that the category in question is that of swans, and that the initial theory is as shown in Figure 12.1 (p. 245). Swanhood is represented as entangled, then, with the internal cause of whiteness, the internal cause of red beaks, and so on. Now imagine (we enter at this point an alternative biological reality) it turns out that these various internal causes are identical: the same thing that determines the color of swans' feathers determines the color of their beaks, imbues them with a tendency to trumpet, and so on.

Imagine also that as further characteristics of swans are discovered, they too turn out to be caused, at root, by precisely the same property or property complex. You will end up with a theory having the structure shown in Figure 15.1, in which there is a single primary explainer of all the characteristic properties, and with which swanhood is entangled. At this point, you are a single arrowhead away from an essentialist theory of swans, that is, a theory that identifies swanhood with the internal cause of swans' characteristic properties. All you need do is hypothesize that the something about swans that causes their characteristic properties is swanhood itself.

As I remarked in Section 11.2 when contemplating the case of gas pressure, you are not deductively obliged to accept this hypothesis, but we humans find the move to be a natural endpoint of something like inference to the best explanation. What best explains the fact that swanhood is entangled with this single cause of all swan-ish properties? That it is identical to the cause. So we turn the arrow of entanglement into the dashed line of definition.

There will surely be other, more complex circumstances under which a category's explanatory status might be upgraded from secondary to primary. Suppose, for example, that all the derived properties in a theory but one have the same cause. You might, again invoking inference to the best

15.4. THE IRREDUCIBILITY OF ALMOST EVERYTHING

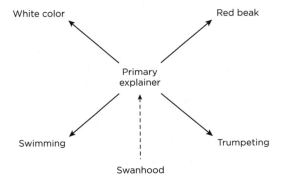

FIGURE 15.1. A theory of swans having a single primary explainer, with which swanhood is entangled.

explanation, identify category and common cause, fingering the category's essential nature as a primary explainer of almost all the derived properties and a secondary explainer of one.

Categories that start out represented by secondary concepts—that start out represented as playing a largely secondary explanatory role—will turn out to be secondary categories if they fail to qualify for the explanatory upgrade (and avoid various other fates, such as abandonment or a downgrade to nonexplanatoriness). Under what circumstances, then, is a category picked out by a secondary concept likely to miss out on an upgrade? What keeps our theory of swans, for example, from positing primary relations between swanhood and any of swans' characteristic properties?

The nub of the answer is that the characteristic properties of swans have causes that are at least partially distinct. Or to put it in terms that make no reference to the swan category: the swan concept is introduced to organize explanatorily (and thus inferentially) a cluster of phenomena that have correlated but substantially distinct causes. When the world in general works this way—when clustered causes, though they may overlap, are rarely identical—there will be relatively few explanatory upgrades.

We have arrived at the conclusion foreshadowed in the opening paragraphs of this chapter: most high-level categories of material things are secondary categories. The final step in my explanation of the irreducibility of almost everything is to give some reason to think that secondary categories are, on the whole, irreducible to the relevant fundamental level, and thus that secondary categories of material things are on the whole irreducible to fundamental physical stuff.

A category is reducible, I wrote above, only if it is possible to specify in fundamental-level terms "what it is to fall into" the category—or in other words, only if it is possible to specify the category's essential nature in fundamental-level terms. It follows immediately that if a category lacks an essential nature, then it cannot be reduced (either to fundamental-level stuff or to any other reduction basis). Secondary categories tend to lack essential natures, and therefore tend to irreducibility.

So why do secondary categories lack essential natures? I can only point to the categories of water and swan as models. With water, you will recall from Section 11.2, the ultimate theory's secondary explanatory structure gives many categorizations using the theory a strong inductive flavor that dissuades an ideal reasoner from thinking of the relevant considerations as constitutively rather than merely evidentially related to the categorical property. This is why the specification of the "water complex," an agglomeration of all the properties that get a specimen classified as water, is rejected as an account of waterhood, along with every other extensionally adequate candidate for theoretical identification. (The dispositional criterion for identifying idemic natures does the rest.)

In the case of swanhood, the lack of an essential nature is due to entanglement's being "looser" than a primary causal relation such as causation. A final theory of swanhood—a theory that takes into account a total set of evidence—attributes to swanhood a secondary explanatory role, characterized by its entanglement with various primary explainers. Several different properties, however, are entangled in this way; each of these plays the role in question, and so no one of them is clearly enough the idemic nature to count, by the dispositional criterion, as such.[11]

For these kinds of reasons, then, categories such as water are irreducible. Although everything material is made of fundamental physical stuff, even

---

[11] The story does not stop there. Many of the role-players may be ruled out as idemic natures because they do not have the correct extension, that is, the extension picked out by the dispositional doctrine of reference. (Precise physical properties may not make the right calls about borderline cases, for example. This phenomenon well illustrates the practical difference between making classifications by using a theory inferentially, and making classifications by counting as a category member anything that "fits" the theory.) The discussion in Section 11.2 shows, however, that sufficiently many candidates remain to prevent us from making a definite theoretical identification.

water and the brains that think about it—although there is nothing to water, or the concept of water, or the thoughts and actions that connect one to the other, but particles doing particular things at particular places and times—waterhood is not a physical thing.

15.5 The Construction of the Secondary World

Constructed—let me consider at last in what sense the secondary categories deserve this philosophical epithet. How, that is, can it make sense to say that waterhood and swanhood are products of the mind, rather than provisions of the world?

For simplicity's sake, let me continue to focus on the material world, though the case of philosophy's secondary categories will require generalization to the epistemic and moral domains among others, which I leave to you. Let me also suppose, for the sake of a bracing metaphysical contrast, that the fundamental-level facts of the material world—facts about the positions and velocities of particles (or about fields or strings) and facts about the primary explanatory relations between them, namely causal influence and nomological dependence—are not in any sense constructed.

Some facts are reducible to the fundamental level, in the sense that what it takes for them to obtain can be defined in terms of fundamental-level facts, but are not themselves fundamental. Centers of mass provide a simple example.[12] Is the property of being a center of mass constructed?

Absolutely not. A center of mass has as its idemic nature a certain configuration of fundamental-level elements of reality, which is to say that it is identical to a certain (complex) fundamental-level state of affairs: for a system to have its center of mass in a certain position is for the positions and masses of its parts to sit in a certain spatiotemporal relation. Thus it has the same ontological status as things at the fundamental level; it is real and not constructed.

This is true in spite of the abstract nature of centers of mass, which makes itself manifest in multiple realizability. A planetary system and a system of two motes of dust may have precisely the same center of

---

[12] Here I am assuming, of course, that mass is a determinate fundamental property, though various aspects of modern physics might suggest otherwise.

mass—they may instantiate the very same property of having a center of mass at such and such a position. Nevertheless, what it is to have this property can be specified in fundamental-level terms, and so is itself simply a high-level property of fundamental material reality—a way that certain parts of fundamental material reality might be.

Such "real definitions," specifying the corresponding entities' idemic natures in fundamental-level terms, can be as complex and abstract as you like without compromising the entities' reality. Suppose that what is to be a desire is to occupy a certain complex causal role with respect to other mental states and to the rest of the physical world.[13] Then, although such a causal role might be realized in many fundamentally different ways—neurons, cogs, transistors, Martian hydraulics—desires are, because definable in fundamental-level terms, compartments of fundamental material reality and so have comparable ontological status. (The fundamental-level specifiability of a causal role requires, note, that the defining causal relations are themselves built from fundamental material constituents, so in this case, that mental causation is a kind of physical causation.)

Call properties that are definable in terms of—hence reducible to—the material world's fundamental level *physical properties*. A physical property, then, is a property with a materially fundamental idemic nature. Physical properties and states of affairs are as real as fundamental-level properties and states of affairs.

The secondary categories are, when irreducible, not physical properties. But in what sense are they conceptual constructs?

Consider a paradigmatic specimen of water. What makes it water and not some other kind of stuff? It is something about the specimen's physical configuration—it is its having a certain, perhaps complex, physical property P. (The relevant property will differ, as I explained in Section 11.3, from specimen to specimen.) This property P is what is called the ground of the specimen's waterhood. What makes it the ground is a "grounding fact": *Anything with P is water*. The difference between real and constructed categories can be found, I suggest, in the provenance of the relevant grounding facts—here borrowing from the recent metaphysical literature on grounding (Fine 2001; Rosen 2010).

---

[13] Never mind that I think this supposition to be false (note 1 of Chapter 14).

15.5. THE CONSTRUCTION OF THE SECONDARY WORLD

Consider a physical property such as having a center of mass at point x. Take some particular system that instantiates this property, that is, that has its center of mass at x. What makes it the case that the system instantiates the property—that it belongs, as it were, to the category of systems with centers of mass at x? Its having a certain physical property P. Of course, P is simply the property of having a center of mass at x. There is no question, then, of where the grounding fact comes from, that is, of why it is the case that *Anything with P has its center of mass at x*. P and the center of mass property are one and the same thing, so the grounding fact has the character of an identity. Nothing substantive is needed to secure the truth of such a fact. It comes "for free."

The case of water is different. Waterhood is not identical to P, and nor does the grounding fact come "for free" in any other way—this because of waterhood's irreducibility to the physical facts.

So what is the origin of the grounding fact? It is not itself a fundamental-level fact, because fundamental-level facts cannot concern nonfundamental matters such as waterhood (Sider 2011, §7.2). But in that case, the grounding fact is nonfundamental, and so its obtaining must be grounded in certain fundamental facts. What could these be?

Philosophers have given several answers,[14] but the correct response, I suggest, is the most traditional: the grounds of grounding are the facts about the concept of water (or better, whatever fundamental facts ground that psychological entity). More specifically—you know how my story goes—they are the facts that decide the extension of the water concept, namely, the facts that decide what we would or would not categorize under the concept if we were to learn a total set of evidence. They include the central aspects of the cognitive structure of the water concept, as well as the facts that determine the truth of the dispositional theory of reference (more psychological facts?[15]) and the facts that determine the validity of our canons of inductive logic. Thus, *we* ground—or more exactly, our thinking about water grounds—the grounding principles in virtue of which there are facts about what counts and what does not count as water.

---

[14] For example, that P itself grounds "Anything with P is water" (deRosset 2013; Bennett 2017).

[15] The question mark indicates my taking out something of a promissory note, or perhaps my taking an Alexandrian sword to the Gordian knot of circularity discussed with more philosophical sensitivity in Section 9.7.

I require one further premise: that there is no waterhood, thus no category of water, without facts about what is and is not water. Since the latter facts are grounded in the concept of water, there is no category of water without a concept of water. Conceptual construction is also categorical construction.

When the gods created the material world, they built everything materially real: they made the fundamental properties, entities, and relations, and so (without any additional labor) they made the physical properties, entities, and relations. Then they went away. There were no basic natural kinds in the world they left behind (though there were physical kinds that we would recognize as approximations to the basic natural kinds).

But there were kind-builders: there were physical things, us, who by developing new inductive theories also constructed new categories. The builders' introjections took them beyond primary explanatory structure, where nature's own categories are indelibly etched, to categories that for diverse purposes bundled together various primary explainers. That is where water, gold, and swanhood came from, to sit alongside but outside of physical reality.

They may be constructed, but the secondary categories are not entirely unreal. They have extensions, and those extensions encompass real objects. Using our secondary concepts we think and say much that is true and significant about the physical world, even about states of affairs that existed before our acts of conceptual construction. Water existed a billion years ago, but neither the concept nor the category of water appeared until geologically recent times. Phlogiston and mermaids may be imaginary, then, but waterhood and swanhood have a certain penumbral substance.

Indeed, the world as we experience it and reason about it and make our way in it is a world largely organized by secondary categories. It is a world of tables and chairs, of leopards and hyraxes and cycad trees, of songs and knives and muffled thumps in the night—all secondary things, I believe, though I have not made that case here. Perhaps much of our philosophical world is secondary also: knowledge like water, "flowing and drawn, and since / our knowledge is historical, flowing, and flown."[16]

---

[16] Bishop, "At the Fishhouses," final lines.

REFERENCES

ACKNOWLEDGMENTS

INDEX

References

Ahn, W. (1998). Why are different features central for natural kinds and artifacts? The role of causal status in determining feature centrality. *Cognition* 69:135–178.

Alexander, J. (2012). *Experimental Philosophy: An Introduction*. Polity, Cambridge.

Alexander, J., and J. M. Weinberg. (2007). Analytic epistemology and experimental philosophy. *Philosophy Compass* 2:56–80.

Anderson, R. L. (2015). *The Poverty of Conceptual Truth: Kant's Analytic/Synthetic Distinction and the Limits of Metaphysics*. Oxford University Press, Oxford.

Atran, S. (1990). *Cognitive Foundations of Natural History*. Cambridge University Press, Cambridge.

Audi, R. (2008). Intuition, inference, and rational disagreement in ethics. *Ethical Theory and Moral Practice* 11:475–492.

Balcerak Jackson, M., and B. Balcerak Jackson. (2012). Understanding and philosophical methodology. *Philosophical Studies* 161:185–205.

Bealer, G. (1996). On the possibility of philosophical knowledge. *Philosophical Perspectives* 10:1–34.

Bealer, G. (1998). Intuition and the autonomy of philosophy. In DePaul and Ramsey (1998), pp. 201–239.

Bennett, K. (2017). *Making Things Up*. Oxford University Press, Oxford.

Berlin, B., D. Breedlove, and P. Raven. (1974). *Principles of Tzeltal Plant Classification*. Academic Press, New York.

Bishop, E. (2011). "At the Fishhouses," in *Poems*, pp. 62–64. Farrar, Straus and Giroux, New York.

Bishop, M. A., and J. D. Trout. (2005). *Epistemology and the Psychology of Human Judgment*. Oxford University Press, New York.

Block, N., and R. Stalnaker. (1999). Conceptual analysis, dualism, and the explanatory gap. *Philosophical Review* 108:1–46.

Boghossian, P. A. (1996). Analyticity reconsidered. *Noûs* 30:360–391.

Bonini, N., D. Osherson, R. Viale, and T. Williamson. (1999). On the psychology of vague predicates. *Mind and Language* 14:377–393.

Boyd, R. (1988). How to be a moral realist. In G. Sayre-McCord (ed.), *Essays on Moral Realism*, pp. 181–228. Cornell University Press, Ithaca, NY.

———. (1999). Homeostasis, species, and higher taxa. In R. A. Wilson (ed.), *Species: New Interdisciplinary Essays*, pp. 141–185. MIT Press, Cambridge, MA.

Brandom, R. (1998). *Making It Explicit: Reasoning, Representing, and Discursive Commitment*. Harvard University Press, Cambridge, MA.

Burgess, A., and D. Plunkett. (2013). Conceptual ethics I. *Philosophy Compass* 8:1091—1101.

Cappelen, H. (2012). *Philosophy without Intuitions*. Oxford University Press, Oxford.

———. (2018). *Fixing Language: An Essay on Conceptual Engineering*. Oxford University Press, Oxford.

Carey, S. (1985). *Conceptual Change in Childhood*. MIT Press, Cambridge, MA.

———. (2009). *The Origin of Concepts*. Oxford University Press, Oxford.

Carnap, R. (1950). *Logical Foundations of Probability*. University of Chicago Press, Chicago.

———. (1952). Meaning postulates. *Philosophical Studies* 3:65–73.

Chalmers, D. J. (1996). *The Conscious Mind*. Oxford University Press, Oxford.

———. (2002). On sense and intension. *Philosophical Perspectives* 18:135–182.

———. (2012). *Constructing the World*. Oxford University Press, Oxford.

———. (2015). Why isn't there more progress in philosophy? *Philosophy* 90:3–31.

Chalmers, D. J., and F. Jackson. (2001). Conceptual analysis and reductive explanation. *Philosophical Review* 110:315–361.

Chomsky, N. (1995). Language and nature. *Mind* 104:1–61.

Chudnoff, E. (2016). *Intuition*. Oxford University Press, Oxford.

Cimpian, A., and E. Salomon. (2014). The inherence heuristic: An intuitive means of making sense of the world, and a potential precursor to psychological essentialism. *Behavioral and Brain Sciences* 37:461–527.

Clement, J. (1983). A conceptual model discussed by Galileo and used intuitively by physics students. In Gentner and Stevens (1983), pp. 325–339.

Coffa, J. A. (1991). *The Semantic Tradition from Kant to Carnap: To the Vienna Station*. Cambridge University Press, Cambridge.

Colaço, D., W. Buckwalter, S. Stich, and E. Machery. (2014). Epistemic intuitions in fake-barn thought experiments. *Episteme* 11:199–212.

Cummins, R. (1998). Reflection on reflective equilibrium. In DePaul and Ramsey (1998), pp. 113–127.

De Cruz, H. (2015). Where do intuitions come from? *Australasian Journal of Philosophy* 93:233–249.

DePaul, M. R., and W. Ramsey (eds.). (1998). *Rethinking Intuition: The Psychology of Intuition and Its Role in Philosophical Inquiry*. Rowman and Littlefield, Lanham, MD.

deRosset, L. (2013). Grounding explanations. *Philosophers' Imprint* 13.7:1–26.

Deutsch, M. (2015). *The Myth of the Intuitive: Experimental Philosophy and the Philosophical Method*. MIT Press, Cambridge, MA.

Devitt, M. (1981). *Designation*. Columbia University Press, New York.

———. (2008). Resurrecting biological essentialism. *Philosophy of Science* 75:344–382.

Devitt, M., and K. Sterelny. (1987). *Language and Reality*. MIT Press, Cambridge, MA.

Field, H. (1973). Theory change and the indeterminacy of reference. *Journal of Philosophy* 70:462–481.

Fine, A. (1975). How to compare theories: Reference and change. *Noûs* 9:17–32.

Fine, K. (2001). The question of realism. *Philosophers' Imprint* 1.2:1–30.

Fodor, J. A. (1975). *The Language of Thought*. Harvard University Press, Cambridge, MA.

———. (1981). The present status of the innateness controversy. In *RePresentations*, pp. 257–316. MIT Press, Cambridge, MA.

———. (1998). *Concepts: Where Cognitive Science Went Wrong*. Oxford University Press, Oxford.

Fodor, J. A., J. D. Fodor, and M. F. Garrett. (1975). The psychological unreality of semantic representations. *Linguistic Inquiry* 6:515–531.

Fodor, J. A., M. F. Garrett, E. C. T. Walker, and C. H. Parkes. (1980). Against definitions. *Cognition* 8:1–105.

Franklin-Hall, L. R. (2014). High-level explanation and the interventionist's 'variables problem'. *British Journal for the Philosophy of Science* 67:553–577.

———. (2016). New mechanistic explanation and the need for explanatory constraints. In K. Aizawa and C. Gillett (eds.), *Scientific Composition and Metaphysical Ground*, pp. 41–74. Palgrave Macmillan, London.

Gelman, S. A. (2003). *The Essential Child: Origins of Essentialism in Everyday Thought*. Oxford University Press, Oxford.

Gendler, T. S. (2000). The puzzle of imaginative resistance. *Journal of Philosophy* 97:55–81.

Gentner, D., and A. Stevens (eds.). (1983). *Mental Models*. Lawrence Erlbaum, Hillsdale, NJ.

Gettier, E. L. (1963). Is justified true belief knowledge? *Analysis* 23:121–123.

Gibbons, J. (2001). Knowledge in action. *Philosophy and Phenomenological Research* 52:579–600.

Goldman, A. (1986). *Epistemology and Cognition.* Harvard University Press, Cambridge, MA.

Goldman, A., and J. Pust. (1998). Philosophical theory and intuitional evidence. In DePaul and Ramsey (1998), pp. 179–197.

Gopnik, A. (1988). Conceptual and semantic development as theory change: The case of object permanence. *Mind and Language* 3:197–216.

Gould, S. J. (1989). *Wonderful Life.* W. W. Norton, New York.

Greco, J. (2010). *Achieving Knowledge: A Virtue-Theoretic Account of Epistemic Normativity.* Cambridge University Press, Cambridge.

Hampton, J. A. (2007). Typicality, graded membership, and vagueness. *Cognitive Science* 31:355–384.

Hampton, J. A., Z. Estes, and S. Simmons. (2007). Metamorphosis: Essence, appearance, and behavior in the categorization of natural kinds. *Memory and Cognition* 35:1785–1800.

Harman, G. (1994). Doubts about conceptual analysis. In M. Michael and J. O'Leary-Hawthorne (eds.), *Philosophy in Mind: The Place of Philosophy in the Study of Mind,* pp. 43–48. Kluwer, Dordrecht.

Haslanger, S. (2000). Gender and race: (What) are they? (What) do we want them to be? *Noûs* 34:31–55.

Hirschfeld, L. (1996). *Race in the Making: Cognition, Culture, and the Child's Construction of Human Kinds.* MIT Press, Cambridge, MA.

Hobbes, T. (1651). *Leviathan, or The Matter, Forme, and Power of a CommonWealth Ecclesiasticall and Civill.* Andrew Crooke, London.

Howson, C. (2001). *Hume's Problem: Induction and the Justification of Belief.* Oxford University Press, Oxford.

Hull, D. (1978). A matter of individuality. *Philosophy of Science* 45:335–360.

Ichikawa, J. J., and M. Steup. (2014). The analysis of knowledge. In E. N. Zalta (ed.), *The Stanford Encyclopedia of Philosophy.* Spring 2014 edition. http://plato.stanford.edu/archives/spr2014/entries/knowledge-analysis/.

Jackman, H. (2003). Charity, self interpretation, and belief. *Journal of Philosophical Research* 28:145–170.

Jackson, F. (1998). *From Metaphysics to Ethics: A Defence of Conceptual Analysis.* Oxford University Press, Oxford.

Jenkins, C. S. (2006). Knowledge and explanation. *Canadian Journal of Philosophy* 36:137–164.

Johnson, M. and J. Nado. (2014). Moderate intuitionism: A metasemantic account. In A. R. Booth and D. P. Rowbottom (eds.), *Intuitions.* Oxford University Press, Oxford.

Johnston, M., and S.-J. Leslie. (2012). Concepts, analysis, generics and the Canberra Plan. *Philosophical Perspectives* 26:113–171.

Joyce, R. (2006). *The Evolution of Morality*. MIT Press, Cambridge, MA.

Kant, I. (1997). *Critique of Pure Reason*. Translated by P. Guyer and A. W. Wood. Cambridge University Press, Cambridge.

Katz, J. J. (1977). The real status of semantic representations. *Linguistic Inquiry* 8:559–584.

Keil, F. C. (1989). *Concepts, Kinds and Conceptual Development*. MIT Press, Cambridge, MA.

Kim, J. (1994). Explanatory knowledge and metaphysical dependence. *Philosophical Issues* 5:51–69.

Knobe, J. (2003). Intentional action and side effects in ordinary language. *Analysis* 63:190–194.

Knobe, J., and B. Fraser. (2008). Causal judgment and moral judgment: Two experiments. In W. Sinnott-Armstrong (ed.), *Moral Psychology*, volume 2, pp. 441–448. MIT Press, Cambridge, MA.

Kornblith, H. (2002). *Knowledge and Its Place in Nature*. Oxford University Press, Oxford.

Kozhevnikov, M., and M. Hegarty. (2001). Impetus beliefs as default heuristics: Dissociation between explicit and implicit knowledge about motion. *Psychonomic Bulletin & Review* 8:439–453.

Kragh, H. (1999). *Quantum Generations: A History of Physics in the Twentieth Century*. Princeton University Press, Princeton, NJ.

Kripke, S. A. (1980). *Naming and Necessity*. Harvard University Press, Cambridge, MA.

———. (1982). *Wittgenstein on Rules and Private Language*. Harvard University Press, Cambridge, MA.

Langford, C. H. (1942). The notion of analysis in Moore's philosophy. In P. A. Schilpp (ed.), *The Philosophy of G. E. Moore*. Open Court, Chicago.

LaPorte, J. (1996). Chemical kind term reference and the discovery of essence. *Noûs* 30:112–132.

Leibniz, G. W. (1981). *New Essays on Human Understanding*. Translated and edited by P. Remnant and J. Bennett. Cambridge University Press, Cambridge.

Lewis, D. (1970). How to define theoretical terms. *Journal of Philosophy* 67:427–446.

———. (1984). Putnam's paradox. *Australasian Journal of Philosophy* 62:221–236.

Locke, J. (1975). *An Essay Concerning Human Understanding*. Edited by P. Nidditch. Oxford University Press, Oxford.

Machery, E. (2017). *Philosophy Within Its Proper Bounds*. Oxford University Press, Oxford.

Machery, E., R. Mallon, S. Nichols, and S. P. Stich. (2004). Semantics, cross-cultural style. *Cognition* 92:B1–B12.

Machery, E., and S. Seppälä. (2011). Against hybrid theories of concepts. *Anthropology and Philosophy* 1:99–127.

Machery, E., S. P. Stich, D. Rose, M. Alai, A. Angelucci, R. Berniūnas, E. E. Buchtel, A. Chatterjee, H. Cheon, I.-R. Cho, D. Cohnitz, F. Cova, V. Dranseika, Á. Eraña Lagos, L. Ghadakpour, M. Grinberg, I. Hannikainen, T. Hashimoto, A. Horowitz, E. Hristova, Y. Jraissati, V. Kadreva, K. Karasawa, H. Kim, Y. Kim, M. Lee, C. Mauro, M. Mizumoto, S. Moruzzi, C. Y. Olivola, J. Ornelas, B. Osimani, C. Romero, A. R. Lopez, M. Sangoi, A. Sereni, S. Songhorian, P. Sousa, N. Struchiner, V. Tripodi, N. Usui, A. V. del Mercado, G. Volpe, H. A. Vosgerichian, X. Zhang, and J. Zhu. (2017a). The Gettier intuition from South America to Asia. *Journal of the Indian Council of Philosophical Research* 34:517–541.

Machery, E., S. P. Stich, D. Rose, A. Chatterjee, K. Karasawa, N. Struchiner, S. Sirker, N. Usui, and T. Hashimoto. (2017b). Gettier across cultures. *Noûs* 51:645–664.

———. (2018). Gettier was framed! In M. Mizumoto, S. P. Stich, and E. McCready (eds.), *Epistemology for the Rest of the World*. Oxford University Press, Oxford.

Major, R. H. (1859). *Early Voyages to Terra Australis, Now Called Australia*. Hakluyt Society, London.

Mallon, R., E. Machery, S. Nichols, and S. P. Stich. (2009). Against arguments from reference. *Philosophy and Phenomenological Research* 79:332–356.

Malmgren, A.-S. (2011). Rationalism and the content of intuitive judgements. *Mind* 120:263–327.

Malt, B. (1994). Water is not H_2O. *Cognitive Psychology* 27:41–70.

Margolis, E. (1998). How to acquire a concept. *Mind and Language* 13:347–369.

Margolis, E., and S. Laurence. (1999). *Concepts: Core Readings*, chap. 1, pp. 3–81. MIT Press, Cambridge, MA.

Mayr, E. (1981). Biological classification: Toward a synthesis of opposing methodologies. *Science* 214:510–516.

McCloskey, M. (1983). Naive theories of motion. In Gentner and Stevens (1983), pp. 299–324.

Medin, D. L., and A. Ortony. (1989). Psychological essentialism. In Vosniadou and Ortony (1989), pp. 179–195.

Medin, D. L., and W. D. Wattenmaker. (1987). Category cohesiveness, theories, and cognitive archeology. In U. Neisser (ed.), *Concepts and Conceptual Development: Ecological and Intellectual Factors in Categorization*. Cambridge University Press, Cambridge.

Murphy, G. L. (1993). Theories and concept formation. In I. V. Mechelen, J. Hampton, R. S. Michalski, and P. Theuns (eds.), *Categories and Concepts: Theoretical Views and Inductive Data Analysis*. Academic Press, Cambridge, MA.

———. (2002). *The Big Book of Concepts*. MIT Press, Cambridge, MA.

Murphy, G. L., and D. L. Medin. (1985). The role of theories in conceptual coherence. *Psychological Review* 92:289–316.

Nagel, J. (2012). Intuitions and experiments: A defense of the case method in epistemology. *Philosophy and Phenomenological Research* 85:495–527.

Nietzsche, F. (1887). *Die fröhliche Wissenschaft*. Second edition. E. W. Fritzsch, Leipzig.

Osherson, D. N., and E. E. Smith. (1981). On the adequacy of prototype theory as a theory of concepts. *Cognition* 9:35–58.

Papineau, D. (2009). The poverty of analysis. *Proceedings of the Aristotelian Society Supplementary Volume* 83:1–30.

Paul, L. A. (2012). Metaphysics as modeling: The handmaiden's tale. *Philosophical Studies* 160:1–29.

Paxton, J. M., and J. D. Greene. (2010). Moral reasoning: Hints and allegations. *Topics in Cognitive Science* 2:511–527.

Peacocke, C. (1992). *A Study of Concepts*. MIT Press, Cambridge, MA.

Prasada, S., and E. M. Dillingham. (2006). Principled and statistical connections in common sense conception. *Cognition* 99:73–112.

Prinz, J. J. (2002). *Furnishing the Mind: Concepts and Their Perceptual Basis*. MIT Press, Cambridge, MA.

Putnam, H. (1962). It ain't necessarily so. *Journal of Philosophy* 59:658–671.

———. (1975). The meaning of 'meaning'. In K. Gunderson (ed.), *Language, Mind and Knowledge*, volume 7 of *Minnesota Studies in the Philosophy of Science*. University of Minnesota Press, Minneapolis.

Quine, W. V. O. (1936). Truth by convention. In O. H. Lee (ed.), *Philosophical Essays for Alfred North Whitehead*. Russell and Russell, New York. Revised version published in W. V. O. Quine, *The Ways of Paradox and Other Essays*, Harvard University Press, Cambridge, MA, 1966.

———. (1951). Two dogmas of empiricism. *Philosophical Review* 60:20–43.

———. (1960). *Word and Object*. MIT Press, Cambridge, MA.

———. (1963). Carnap and logical truth. In P. A. Schilpp (ed.), *The Philosophy of Rudolf Carnap*. Open Court, Chicago.

———. (1969). Natural kinds. In *Ontological Relativity and Other Essays*, pp. 114–138. Columbia University Press, New York.

Ramsey, W. (1998). Prototypes and conceptual analysis. In DePaul and Ramsey (1998), pp. 161–177.

Rawls, J. (1999). *A Theory of Justice*. Revised edition. Harvard University Press, Cambridge, MA.

Reichenbach, H. (1951). *The Rise of Scientific Philosophy*. University of California Press, Berkeley, CA.

Rini, R. A. (2015). How not to test for philosophical expertise. *Synthese* 192:431–452.

Rips, L. J. (1989). Similarity, typicality, and categorization. In Vosniadou and Ortony (1989), pp. 21–59.

———. (1995). The current status of research on concept combination. *Mind and Language* 10:72–104.

———. (2001). Necessity and natural categories. *Psychological Bulletin* 127:827–852.

Rosch, E. (1978). Principles of categorization. In E. Rosch and B. Lloyd (eds.), *Cognition and Categorization*, pp. 27–48. Lawrence Erlbaum, Hillsdale, NJ.

Rosch, E., and C. B. Mervis. (1975). Family resemblances: Studies in the internal structure of categories. *Cognitive Psychology* 7:573–605.

Rosen, G. (2010). Metaphysical dependence: Grounding and reduction. In B. Hale and A. Hoffman (eds.), *Modality: Metaphysics, Logic, and Epistemology*, pp. 109–135. Oxford University Press, Oxford.

Salmon, W. C. (1997). Causality and explanation: A reply to two critiques. *Philosophy of Science* 64:461–477.

Shope, R. K. (1983). *The Analysis of Knowing*. Princeton University Press, Princeton, NJ.

Sider, T. (2011). *Writing the Book of the World*. Oxford University Press, Oxford.

Siegel, S. (2012). *The Contents of Visual Experience*. Oxford University Press, Oxford.

Singer, P. (1981). *The Expanding Circle: Ethics, Evolution, and Moral Progress*. Farrar, Straus, & Giroux, New York.

Smith, E., and D. Medin. (1981). *Categories and Concepts*. Harvard University Press, Cambridge, MA.

Sosa, E. (2007). *A Virtue Epistemology*. Oxford University Press, Oxford.

Sosa, E. (2009). A defense of the use of intuitions in philosophy. In D. Murphy and M. Bishop (eds.), *Stich and His Critics*, pp. 101–112. Wiley-Blackwell, Malden, MA.

Sterelny, K. (1983). Natural kind terms. *Pacific Philosophical Quarterly* 64:110–125.

Stich, S. P. (1990). *The Fragmentation of Reason*. MIT Press, Cambridge, MA.

———. (1992). What is a theory of mental representation? *Mind* 101:243–261.

———. (1993). Moral philosophy and mental representation. In M. Hechter, L. Nadel, and R. E. Michod (eds.), *The Origin of Values*, pp. 215–228. Aldine de Gruyter, New York.

Stich, S. P., and K. P. Tobia. (2016). Experimental philosophy and the philosophical tradition. In J. Sytsma and W. Buckwalter (2016), pp. 5–21.

Stoljar, D. (2017). *Philosophical Progress: In Defence of a Reasonable Optimism*. Oxford University Press, Oxford.

Street, S. (2006). A Darwinian dilemma for realist theories of value. *Philosophical Studies* 127:109–166.

Strevens, M. (2000). The essentialist aspect of naive theories. *Cognition* 74:149–175.

———. (2004). Bayesian confirmation theory: Inductive logic or mere inductive framework? *Synthese* 141:365–379.

———. (2007). Why represent causal relations? In A. Gopnik and L. Schulz (eds.), *Causal Learning: Psychology, Philosophy, Computation*, pp. 245–260. Oxford University Press, New York.

———. (2008a). *Depth: An Account of Scientific Explanation*. Harvard University Press, Cambridge, MA.

———. (2008b). Physically contingent laws and counterfactual support. *Philosophers' Imprint* 8.8:1–20.

———. (2011). Probability out of determinism. In C. Beisbart and S. Hartmann (eds.), *Probabilities in Physics*, pp. 339–364. Oxford University Press, Oxford.

———. (2012a). Ceteris paribus hedges: Causal voodoo that works. *Journal of Philosophy* 109:652–675.

———. (2012b). The explanatory role of irreducible properties. *Noûs* 46:754–780.

———. (2012c). Theoretical terms without analytic truths. *Philosophical Studies* 160:167–190.

———. (2014). High-level exceptions explained. *Erkenntnis* 79:1819–1832.

Swain, S., J. Alexander, and J. M. Weinberg. (2008). The instability of philosophical intuitions: Running hot and cold on Truetemp. *Philosophy and Phenomenological Research* 76:138–155.

Sytsma, J., and W. Buckwalter (eds.). (2016). *A Companion to Experimental Philosophy*. Wiley-Blackwell, Hoboken, NJ.

Turri, J. (2016). Knowledge judgments in "Gettier" cases. In J. Sytsma and W. Buckwalter (2016), pp. 337–348.

Turri, J., W. Buckwalter, and P. Blouw. (2015). Knowledge and luck. *Psychonomic Bulletin & Review* 22:378–390.

Unger, P. (1979). *Ignorance: A Case for Scepticism*. Oxford University Press, Oxford.

———. (2014). *Empty Ideas: A Critique of Analytic Philosophy*. Oxford University Press, Oxford.

Valdesolo, P., and D. DeSteno. (2006). Manipulations of emotional context shape moral judgment. *Psychological Science* 17:476–477.

Vosniadou, S., and A. Ortony (eds.). (1989). *Similarity and Analogical Reasoning*. Cambridge University Press, Cambridge.

Weatherson, B. (2003). What good are counterexamples? *Philosophical Studies* 115:1–31.

Wedgwood, R. (2018). The internalist virtue theory of knowledge. *Synthese*. https://doi.org/10.1007/s11229-018-1707-x.

White, R. (2005). Epistemic permissiveness. *Philosophical Perspectives* 19:445–459.

Williamson, T. (1994). *Vagueness*. Routledge, London.
———. (2000). *Knowledge and Its Limits*. Oxford University Press, Oxford.
———. (2007). *The Philosophy of Philosophy*. Blackwell, Oxford.
———. (2011). Philosophical expertise and the burden of proof. *Metaphilosophy* 42:215–229.
Wilson, M. (1982). Predicate meets property. *Philosophical Review* 91:549–589.
Wright, J. C. (2010). On intuitional stability: The clear, the strong, and the paradigmatic. *Cognition* 115:491–503.

Acknowledgments

This project started out as a draft of several chapters for a possible but non-actual Ph.D. thesis at Rutgers University, written in late 1993 and early 1994. It was Steve Stich who inspired both the question and the approach. I don't suppose it turned out quite as he would have wished, but then the contrarian's true disciple is the one who denies him. Many thanks to Steve for his patient mentoring during those years.

Numerous people have commented on the evolving manuscript since that time: Mike Bishop, Dave Chalmers, Henry Jackman, Krista Lawlor, Matt Lindauer, Edouard Machery, Marco Nathan, Laurie Paul, Gina Rini, Brian Weatherson, the participants in a Corridor reading group in 2009, the audience at a Rutgers workshop on philosophical methodology in 2010, the audience at a *Mind and Language* seminar at NYU in 2013, and the attendees at my NYU seminar on philosophical methodology in 2016. My thanks to everyone for comments and advice at all scales (and apologies to those inadvertently left off the list).

Thanks also to the John Templeton Foundation for their support as part of the *Varieties of Understanding* project.

Finally, let me say how much I appreciate the editorial guidance of Lindsay Waters at Harvard University Press over the years, along with the support of others at the Press and working on the book's production, especially Paul Anagnostopoulos at Windfall Software.

Opening and concluding epigraph credit: Excerpt from "At the Fish-houses" from POEMS by Elizabeth Bishop, copyright © 2011 by The Allice H. Methfessel Trust. Published by Farrar, Straus and Giroux in North America and Jonathan Cape in the United Kingdom and Commonwealth. Reprinted by permission of Farrar, Straus and Giroux and The Random House Group Limited.

Index

a priori knowledge, 7, 34–35, 125–130, 131, 180
Ahn, Woo-kyoung, 82
Alexander, Joshua, 15, 72, 199
analytic truth, 26, 79
　analytic/synthetic distinction, 44–45, 112
　consequence of conceptual containment, 24
　consequence of implicit definition, 28
　rare, 43–47, 51
Anderson, Lanier, 23n1
Anomalocaris, 152
Atran, Scott, 75n1, 188n2
Audi, Robert, 68

Balcerak-Jackson, Magdalena & Brendan, 147
Bayesian epistemology. *See* probabilistic epistemology
Bealer, George, 8n4, 68
behaviorism, 39n8
Bennett, Karen, 323n14
Berlin, Brent, 75n1, 188n2
Bishop, Elizabeth, 114, 324
Bishop, Michael, 229
Block, Ned, 127
Boghossian, Paul, 34, 44
Bolzano, Bernard, 25–26

Bonini, Nicolao, 311n8
Boyd, Richard, 155, 191n4, 282
Brandom, Robert, 66
Burgess, Alexis, 229

Canberra Plan. *See under* philosophical analysis, vindication
Cappelen, Herman, 68, 229
Carey, Susan, 65, 83, 303
Carnap, Rudolf, 26, 42n11, 204
case certainty, 5–6, 210
　explained by inductive analysis, 139, 274, 280–281, 283, 284–287
　explained by intensional analysis, 50, 55–56, 120
　explained by modern analysis, 31–32, 37–38, 46–47
　implies conceptual closure, 8, 273
　not certain, 6, 281
　skepticism about, 7, 56, 73
　weak versus strong explanations, 6, 139–140
case judgments, 3, 67–71. *See also* case certainty; philosophical analysis, vindication
　deductive, 32, 109, 140n3
　diverse, 71–73, 73n3, 187, 197–198, 200–202, 204–207
　edge cases, 45, 240, 247–249, 258, 262

case judgments *(continued)*
  expert versus novice, 199–200, 209–210, 287
  inductive, 101–102, 138–139, 204–206, 310
  scope, 7
  social engineering of, 73, 200, 209
  as type 1½ reasoning. *See* reasoning, type 1½
  uncertain, 167–168, 203, 310–311, 313–316
  unreliability. *See under* philosophical analysis, skepticism
categorization. *See also* case judgments
  graded membership, 309–310
  typicality effects, 78, 82, 86n8
Chalmers, David, 12n6, 181n23, 258
  complexity of essential natures, 55n3, 195
  vindication of analysis, 125–130, 162
Chomsky, Noam, 189
Chudnoff, Elijah, 68
Cimpian, Andrei, 93
Clement, John, 144
Coffa, Alberto, 23n1, 26n4
cognitive conceptual structure. *See* concept, token and cognitive structure
Colaço, David, 205n13
concept. *See also* concept, theories of; concept, types of
  compositional, 23, 41, 76, 89–90, 212–213
  empty. *See* reference, failure
  open versus closed. *See* substantiality, conceptual openness and
  primitive/atomic, 27, 29, 90, 213
  secondary, 20, 296, 319, 324. *See also* secondary category
  semantic versus psychological, 26–27, 28, 30, 45–46, 65
  token and cognitive structure, 41–42, 64–65, 106, 201
concept acquisition, 89, 106, 212, 297–304. *See also* introjection

Quinean bootstrapping, 303–304
concepts, specific
  chemical substance, 282–283
  justification, 265–269, 284
  knowledge, 71–72, 114–115, 264–265, 284, 286
  number, 65
  reference, 125, 228
concepts, theories of
  causal minimalism, 97–101, 103–106, 233, 235–236, 245, 302
  classical, 38–43, 44, 64, 77, 112
  exemplar, 78n4
  inductivism, 17, 106–112, 116, 123, 147, 162, 202–203, 213, 257. *See also* inductivism, appeals to semantics
  molecular, 23–24, 26–27, 28, 41, 52, 76, 90
  prototype, 64, 78–82, 85, 89, 245. *See also under* philosophical analysis
  psychological essentialism, 83–96, 97, 99–101, 104
  theory-theory, 83, 86, 109–111
concepts, types of
  basic natural kind, 16–17, 75, 104, 113, 236, 238–239
  culinary, 232, 235–236
  epistemic. *See under* concepts, specific
  folk genus, 75, 87n9, 92, 188, 233
  mental state, 277, 286, 291
  philosophical, 16–17, 114–116, 136, 254, 273
conceptual analysis. *See also under* philosophical analysis, vindication
  classical, 23–25, 29, 32, 133
  hypothetical, 56–61, 117–118, 133
  intensional, 49–51, 54, 118–122
  modern, 25–38, 39, 45–47, 51, 79, 117, 119n1, 133, 135–136, 137
conceptual change, 65. *See also* reference, change in
conceptual complexity, 55n3, 194–197, 292–293
conceptual containment. *See* concepts, theories of, molecular

conceptual diversity, 71–72, 197–198, 228
conceptual engineering, 229, 261
conceptual role semantics. *See* inferential role semantics
conceptual stability, 105–106, 201
conceptual structure. *See* concept, token and cognitive structure
conceptual truth. *See* analytic truth
constructivism, 40, 275, 321–324. *See also under* philosophical analysis, vindication
core causal belief, 85, 97, 242. *See also* entanglement; explanation, starburst structure
    causal-explanatory individuation, 173–174, 219, 234, 267–268
    cognitive significance, 98–99, 174–175, 178, 247–248
    corrigibility, 87–88, 104
corollary to causal minimalism, 101–105
Cummins, Robert, 145

days of the week, 27, 43n13
De Cruz, Helen, 68
definition. *See also* analytic truth; concepts, theories of, classical
    implicit, 27, 29, 42
    real versus nominal, 25, 27–28, 34, 322
derived property, 253, 314
deRosset, Louis, 323n14
Descartes, René, 13, 25
Deutsch, Max, 69n1
Devitt, Michael, 92, 153
discovery scenario (Kripke etc.), 81, 87–88, 90–92, 100, 151–153, 169–173
dispositional approach to reference, 155–156, 207–209
    borderline cases, 168–169, 207, 208–209, 312–313
    doctrine versus account, 181–182, 222
    Socratic oracle, 156–157, 160
    total set of evidence, 155, 157, 159–160
division of linguistic labor, 178–179

edge cases. *See under* case judgments
empirical indifference, 271–273
    functionally explained, 276–277, 281, 283–289
    topically explained, 274–276
empirically informed philosophy. *See under* philosophical analysis
entanglement, 98–99, 99n2, 206, 242–244, 248, 268n10, 320
    cognitive role, 245–247, 252, 253, 258
    explanatory role, 252
epistemic denaturing, 217, 219–220
essence postulate, 57. *See also* conceptual analysis, hypothetical
essential nature
    complicated, 55, 186–187, 195, 218
    explanatory, 221
    extensional, 224–225
    idemic, 214, 216, 321–322
    rare, 211, 225, 320
    target of analysis, 4, 226
essentialism
    biological, 92, 101, 104
    psychological. *See under* concepts, theories of
expanding the circle, 256
experimental philosophy, 71–74, 161n13, 187, 197–199, 228
    negative program, 15, 71, 127
expertise defense. *See* case judgments, expert versus novice
explanation. *See also* starburst explanatory structure
    entanglement, explanatory role of, 252
    holistic scheme, 278, 284, 288, 291–293
    justificatory, 263–264, 266, 268–269
    philosophical, 231, 251–252, 254, 263, 284, 288
    primary versus secondary, 251–252, 254, 317, 320
    relativism, 235

Field, Hartry, 157
Fodor, Jerry, 66, 110

Fodor, Jerry *(continued)*
  against classical concepts, 40–41, 42–43, 44
  atomicity of concepts, 90
  conceptual nativism, 106, 300
  folk theory. *See* naive theory
Franklin-Hall, Laura, 224n9
Frege, Gottlob, 25–26, 34
front-loading argument, 127–129

Gelman, Susan, 83, 86
Gendler, Tamar, 289n8
generative semantics, 43
Gettier cases, 7, 72, 270
  case certainty of, 6, 6n2, 37, 73, 284–285
  explained, 205, 265, 285–286
Gibbons, John, 265n9
Gödel cases, 71–72, 198–199, 228
Goldman, Alvin, 4n1, 266
Gopnik, Alison, 83
Greco, John, 265n8
Greene, Joshua, 201
grounding, 27n6, 29, 322–323

Hampton, James, 81n5, 309
Harman, Gilbert, 44
Haslanger, Sally, 229
Heisenberg, Werner, 260n4
Hirschfeld, Lawrence, 232
Hobbes, Thomas, 30
Hull, Clark, 39n8
Hull, David, 192n5

implicit definition. *See under* definition
inductive analysis, 134–138, 141, 196–197, 213–214
inductive plasticity, 204–207, 208–209
inductivism, appeals to semantics, 179, 307–308, 312
inferential role semantics, 110–111, 302
inferentialism, 53, 66, 130–132
insect, 196–197
introjection, 301–303, 317
  of concept of knowledge, 303
  Margolis module, 300–301

intuitions, 3, 14, 68, 71. *See also* case judgments

Jackman, Henry, 155n4
Jackson, Frank, 14n7, 30, 152n1
Jenkins, C. S., 265n8
Johnson, Michael, 155
Johnston, Mark, 46
Joyce, Richard, 143, 274

Kamprad, Ingvar, 1
Kant, Immanuel, 23n1, 25, 34
Katz, Jerrold, 41n10
Keil, Frank, 68, 69, 80–81, 87n9
Kim, Jaegwon, 251
Knobe, Joshua, 71
Kornblith, Hilary, 17n8, 35, 266
Kozhevnikov, Maria, 144
Kripke, Saul, 34, 81, 88–89, 153, 215n3
Kripkean discoveries. *See* discovery scenario (Kripke, etc.)

Langford, C. H., 25
language of thought. *See* representational theory of mind
LaPorte, Joseph, 189
Laurence, Stephen, 38–39
Leibniz, G. W., 23, 300
Leslie, Sarah-Jane, 46
Lewis, David, 124, 146, 181
linguistic citizenship, 179, 208
Linnaean taxonomy, 75n1, 188, 191, 197n9
Locke, John, 23, 25, 76
logical positivism, 26

Machery, Edouard, 68, 73n3
  against philosophical analysis, 15, 127
  experimental philosophy, 71, 72, 270n11
  parochialism argument, 228–229
  unreliability argument, 198–199, 200–201, 202
Mallon, Ron, 182n24
Malmgren, Anna-Sara, 7
Malt, Barbara, 93–94, 189

Margolis module. *See under* introjection
Margolis, Eric, 38–39, 43, 300
Mayr, Ernst, 191
McCloskey, Michael, 144
mechanism-generic hypothesis, 267–268, 278, 284, 288
Medin, Douglas, 39, 83, 93, 110
meter stick, 34
method of cases. *See* philosophical analysis
molecular duplicate (of organism), 192–193, 205–206, 209, 247
Moore, G. E., 26
Murphy, Gregory, 39n8, 81, 83, 110
myth of depth, 113–114, 138–139, 311n8

Nado, Jennifer, 155
Nagel, Jennifer, 102n3
naive theory, 84–85, 188, 290
  biology, 75n1
  chemistry, 282–283
  epistemology, 270n11, 285
  ethics, 275
  physics, 143–144
  psychology, 277–280
  semantics, 106, 125, 308, 312
Nietzsche, Friedrich, 146
Nix v. Hedden, 232

ordinary belief, 107, 108–109, 112, 124, 138n2
Ortony, Andrew, 83
Osherson, Daniel, 82

Papineau, David, 15
paradox of analysis, 24, 32
Paul, L. A., 15
Paxton, J. M., 201
Peacocke, Christopher, 53
philosophical analysis, 2–3. *See also* conceptual analysis; inductive analysis; philosophical analysis, skepticism; philosophical analysis, vindication
  empirically informed, 10–13, 255–256, 271–272
  of empty categories, 259–262, 291–293
  failure, 185–187, 259
  not about concepts, 35–38, 137–138
  prototype theory and, 40, 79
  of secondary categories, 238–240, 257–258, 319–321
philosophical analysis, skepticism, 185–187
  insubstantiality of philosophical categories, 7–9, 228–229
  no essential natures, 212, 225
  unreliability of case judgments, 198–202
philosophical analysis, vindication
  Canberra Plan, 14, 124
  as conceptual analysis, 30–33
  constructivist, 14, 27–28, 30, 34, 59, 118, 146–147
  empirical (Darwinian), 13, 143–145
  as inductive analysis, 137–138. *See also* reflexivity of reference; substantiality
  nativist, 13–14, 59, 145–146
  reference-first, 53–55, 123, 162
  scientistic, 14–15
  success conditions, 9–10
philosophical analyst, working versus self-conscious, 35–37, 55–56, 119–120, 136, 285
philosophical category, 4
philosophical knowledge, 1–2, 226
  domains, 229–230
  as explanatory knowledge, 231, 257–258, 270
  of necessary truth, 7, 178
philosophical progress, 258–259
Plato, 13
Plunkett, David, 229
Prasada, Sandeep, 252
primary category, 254
prime number, 32, 44, 112
Prinz, Jesse, 42
probabilistic epistemology, 67, 168n16, 204, 305–306
Putnam, Hilary, 81, 88–89, 173

Quine, Willard Van Orman, 15, 44–45, 112, 122, 239
Quinean bootstrapping, 303–304

Ramsey, Frank, 124
Ramsey, William, 40, 79
Rawls, John, 57–58
Rawlsian veil, 32, 54–55, 57, 59, 147
  case certainty and, 38, 47, 55–56
  explains apparent rarity of analyticity, 46–47
  reverse, 102–103
reasoning
  inference to the best explanation, 85–86, 206, 215, 318–319
  probabilistic. *See* probabilistic epistemology
  type 1 and 2, 68
  type 1½, 70, 103
reducibility and irreducibility, 316, 319–321
reference. *See also* dispositional approach to reference; reflexivity of reference
  causal-historical theory, 49–50, 153, 170, 177
  change in, 157–158, 196–197
  failure, 157, 163, 259–262, 280, 304–308
  indeterminacy, 157, 196–197
  magnetism, 181
  "qua" problem, 121
  Ramsey-Lewis theory, 124, 146, 151–153, 195
reference-first strategy. *See under* philosophical analysis, vindication
reflective equilibrium, 57–58
reflexivity of reference, 146–147, 149–154, 163, 182, 294. *See also* dispositional approach to reference
  inductivism and, 147
  principle of charity, 149, 153–154
  substantiality and, 147–148, 150–151, 227
Reichenbach, Hans, 9
representational theory of mind, 66

Rini, Regina, 199
Rips, Lance, 42n10, 81, 82
Rosch, Eleanor, 78
Russell, Bertrand, 26

Salmon, Wesley, 175
schwann, 87–88, 151–152, 153, 169–173, 306
secondary category, 254, 295, 296. *See also under* philosophical analysis
  common, 295, 324
  constructed, 322–324
  insubstantial, 258
  irreducible, 319–321
  vague, 313–316
self-conscious philosophical analyst. *See* philosophical analyst, working versus self-conscious
Seppälä, Selja, 201
sexual dimorphism, 81, 152
Shope, Robert, 186
Sider, Ted, 194–195, 323
Siegel, Susanna, 70
Singer, Peter, 256
Smith, Edward, 39, 82
Sosa, Ernest, 73n3, 265n8
Stalnaker, Robert, 127
starburst explanatory structure, 84, 91–92, 103–104, 252–254, 282–283
starter concept. *See* naive theory
Sterelny, Kim, 177n21
Stich, Stephen, 15, 40, 79, 143n4, 185, 228
stipulativity, 52–54, 118, 121, 122, 123, 125
Stoljar, Daniel, 258
Street, Sharon, 143, 274
substantiality, 1
  conceptual openness and, 7–9, 90, 150–151, 234–235, 256–257, 289–290
  explanatory characterization, 231, 255
  of natural kinds, 16, 153, 234–235
Swain, Stacey, 72, 199
swan, 220, 318–319. *See also* schwann

analysis, 190–193
black, 77, 80, 166–167, 192
psychology, 76, 164–167, 241, 310–311, 314–316
system 1 and 2. *See* reasoning, type 1 and 2

token conceptual structure. *See* concept, token and cognitive structure
total set of evidence. *See under* dispositional approach to reference
transformation experiments (Keil), 80–81, 86–87, 100, 171, 210, 247
Trout, J. D., 229
Truetemp, 72, 73, 199
Turri, John, 73, 200, 205, 209, 286
Twin Earth, 173
type 1, 2, and 1½ reasoning. *See under* reasoning

Unger, Peter, 2, 260n6

vagueness, 169, 308–316
  chemical sorites, 313
  evolutionary sorites, 310–311, 314–316
Valdesolo, Piercarlo, 72

water, 101, 159–160, 216–220, 313–314
  analysis, 11, 188–190, 238, 239
  psychology, 93–96, 173–177
water complex, 217–218
Weatherson, Brian, 58n4, 59, 144n6, 152n1
Wedgwood, Ralph, 265n8
Weinberg, Jonathan, 72, 199
White, Roger, 204n12
Williamson, Timothy, 7, 145n7, 186n1, 199, 265n9
  against analytic truth, 44–45
  against conceptual analysis, 35, 36–37
  against philosophical analysis, 185, 212
  against reflective equilibrium, 58n4
  armchair philosophy, 136, 149
  vagueness, 309
Wilson, Mark, 157
Wolff, Christian, 23
working philosophical analyst. *See* philosophical analyst, working versus self-conscious
world-to-category conditional, 125–130, 131, 162, 274
world-to-essential-nature conditional, 11–12, 272
Wright, Jennifer, 73n3, 199, 203